Lecture Notes in Earth Sciences

Lecture Notes in Earth Sciences

Edited by Somdev Bhattacharji, Gerald M. Friedman,
Horst J. Neugebauer and Adolf Seilacher

13

Tadeusz M. Peryt (Ed.)

Evaporite Basins

Springer-Verlag
Berlin Heidelberg GmbH

Editor

Dr. Tadeusz M. Peryt
Instytut Geologiczny
ul. Rakowiecka 4, PL-00-975 Warszawa, Poland

ISBN 978-3-540-18679-3   ISBN 978-3-540-48069-3 (eBook)
DOI 10.1007/978-3-540-48069-3

2132/3140-543210

"Interpretation of evaporites is still very much an art and many
stratigraphic units have been interpreted in very different ways.
... Evaporite sedimentology is in a considerable state of flux
and probably will remain so for some years to come."

A. C. Kendall 1978 'subaqueous evaporites'

Preface

Contemporary evaporites are principally to be found in three environments: sabkha,
salina, and hypersaline lake, which have also been identified in ancient evaporites;
however, for the interpretation of any particular evaporite basin often contrasted
sedimentary models are proposed. The concept of evaporite drawdown and deposition
of evaporites in great desiccated basins led to a debate on the origin of salt giants,
the conclusion to which was that evaporites may form or deposit in a wide spectrum
of environments from continental sabkha to deep basin. For a number of reasons, the
definite models of evaporite deposition have been not yet formulated, one reason
being that in fact only a few evaporite basins have been studied in detail (Miocene
Mediterranean basin, the Zechstein basin of Central and NW Europe, and the Upper
Silurian basin of Michigan being probably the best-known cases).
The second important factor is that evaporites evoluted in the history of the earth
and our understanding of the evolution is very imperfect.

These two reasons led to the conclusion manifested in the theme of this volume, that
it would be desirable to summarize our knowledge on some less-known evaporite basins
such as those located in China (some of which, like the Tarim Basin discussed in this
volume, foreign visitors are not allowed to enter) or European evaporite basins known
from boreholes.
Many potential authors, enthusiastic at the beginning of the work on the volume,
could not finish their papers within the promised time, and the delay in publishing
occasioned by late chapters would have been - and already was - detrimental to those
authors who had completed their papers in time. The unintended bias does not, we hope,
affect the main message of this volume.

Michal Pawlik and Agnieszka Siara are thanked for technical assistance.

Tadeusz Marek Peryt

Table of Contents

Introduction

Evaporites may form in a spectrum of environments from continental sabkha (playa) to
deep basins (see Kendall 1978 a, b, Schreiber 1978, 1986, Friedman and Krumbein
1985, for review). In the last two decades, many ancient evaporite basins have been
interpreted using the sabkha model and the deep desiccated basin model, the former
not excluding the latter. However, growing evidence has been gathered indicating that
most evaporites are formed in subaqueous environments, so that it cannot be reasonably
expected that one depositional model alone will explain the entire basin fill.

The chapters in this volume discuss characteristic examples of evaporite basins,
mostly of moderate size. Aspects of a saline giant, the Zechstein basin of Central
and NW Europe, have been considered in Volume 10 of "Lecture Notes in Earth Sciences".

Muir presents a set of facies models for the Precambrian evaporites of Australia,
ranging from continental alkaline playas through continental sabkhas to barred basins.
Desiccated deep marine basins have been not recognized in the Precambrian of Australia. Xu Xiao Song discusses the Sinian evaporites of southern Sichuan, China which
form the top part of the carbonate platform development. Evaporites are related to
two models: The Lagoon-salt lake model and the sabkha-salt lake model. During the late
stage of evaporite deposition sea level fall caused the salt lake in the sabkha to
evolve into a continental salt lake. Rouchy et al. describe Visean evaporites along
the Faille du Midi overthrust in northern France through Belgium to the Netherlands.
Anhydrites are intercalated in limestones (dolomites are rare), and the faunal record
shows a progressive restriction from a marine environment and a marine origin of the
brines which generated the sulfate interbeds. The evaporitic stage was related to the
fall in sea level. On the evidence of abundant relics and pseudomorphs of gypsum, it
is assumed that subaqueous gypsum was an important initial deposit. The predominant
nodular and mosaic structures are interpreted as resulting from burial conversion of
gypsum to anhydrite. Additionally, primary structures have been destroyed due the
Hercynian deformations, and the resulting structures mimic sedimentary structures.
This, however, does not imply that even in such tectonized sequences it is impossible
to interpret sedimentary facies. Rouchy et al. also discuss the origin of the Great
Visean Breccia and conclude that it formed by collapse after dissolution of anhydrite
(and possibly salt) interbeds. In fact, halite is not known up to now from these evaporites, but the presence of recycled halite in the Rotliegendes of the North Sea and
NW Germany might be related to the extensive dissolution of halite which began with
the Permian denudation.

Lecture Notes in Earth Sciences, Vol. 13
T. M. Peryt (Ed.), Evaporite Basins
© Springer-Verlag Berlin Heidelberg 1987

The evaporites described by Wu Yinglin et al. are included into two depositional models. The first one,the platform sabkha model, bears analogies with the recent sabkhas of the Persian Gulf although the topography of the platform has been, as is supposed, more diversified than the recent topography in Abu Dhabi. There existed some distinct depressions in the margin of the platform, forming coastal lakes, and in the inner parts of the platform. Accordingly, sabkha and saline environments pass laterally, or transform, one into another. It was possible to distinguish three stages in the evolution of platform sabkha: the coastal salt lake (salt pan) stage when mainly halite has been deposited, the sabkha stage (when gypsum, halite, and polyhalite have been deposited), and the playa lake stage (when deposition of anhydrite and polyhalite dominated). The second model, the desiccation-lagoon model refers to the salt lake separated from the lagoon in the platform by a rapid regression.

Geisler-Cussey discusses the effects of continental influences on evaporitic sedimentation in a part of the Middle Muschelkalk salt basin. This basin exemplifies a quite common geological situation when the intracontinental basin had marine characteristics but a continental context, a tendency more accentuated in the Keuper basin. The sedimentary record in the Middle Muschelkalk indicates a cyclic evolution of salinity which was related both to the nature of the connection with the Tethyan water supply through the Silesian Sill, that allowed considerable brine concentration before reaching the Paris Basin, and to the freshwater supply from the continental. The basin was not very deep and the brines were not commonly stratified. Except for the centre of the basin, the salts have been affected by dissolution—reprecipitation processes, and the salt has been recycled both by freshwater and unsaturated seawater, the latter probably being of major significance. Qiu Dongzhou characterizes the deposition in a continental lake basin which has been occasionally invaded by the sea, and hence possesses lake features with a few marine characters; such a complex origin is rarely considered in the fossil record.

Monty et al. discuss a much disputed problem of reef-evaporite relations: the answer is of great geological end economical significance. They demonstrate that the origin of reef complexes in the Red Sea Miocene predated the deposition of massive evaporites. Most evaporites formed subaqueously during lowstands in sea level, although this does not implicate overall desiccation of the basin. Sea level fluctuations resulted in alternated phases of exposure and flooding of reefs, and hence, their complex diagenesis and the prolific development of stromatolites. The chapter is a standard for studies of evaporite-related stromatolites as well.

The volume shows that evaporitic basins have been dynamic systems, and that sea level changes have been of important significance for evaporite deposition: hitherto, their importance has been recognized for the pre-evaporitic phase (e.g. Peryt 1984 for the Zechstein basin).

Tadeusz Marek Peryt

References

Friedman G M, Krumbein W E (Eds) (1985) Hypersaline ecosystems - The Gavish Sabkha. Ecological Studies, Springer-Verlag Berlin Heidelberg New York, 53: 484 pp
Kendall A C (1978 a) Facies models 11. Continental and supratidal sabkha evaporites. Geosci Canada, 5, pp 66-78
Kedall A C (1978 b) Facies models 12. Subaqueous evaporites. Geosci Canada, 5, pp 124-139
Peryt T M (1984) Sedymentacja i wczesna diageneza utworów wapienia cechsztyńskiego w Polsce zachodniej. Prace Inst Geol 109: 80 pp
Schreiber B C (1978) Environments of subaqueous gypsum deposition. SEPM Short Course, 4, pp 43-73
Schreiber B C (1986) Arid shorelines and evaporites. In: Reading, H.G. (ed.), Sedimentary environments and facies, pp 189-228. Blackwell, Oxford

FACIES MODELS FOR AUSTRALIAN PRECAMBRIAN EVAPORITES

M.D. Muir

CRA Exploration Pty. Ltd.

P.O. Box 656

Fyshwick, ACT, Australia 2609

INTRODUCTION

There are records of abundant evaporites in the Archean and Proterozoic of the Australian Subcontinent. The majority of the evaporite crystals have been replaced by a variety of other minerals, but sufficient evidence remains to indicate extensive developments of evaporitic conditions at different times in the Australian Precambrian. Few of the evaporite occurrences have been fully studied, but facies models have been developed for a number of the evaporitic sequences (Figure 1).

For the purposes of this paper, only better known, highly evaporitic sequences will by discussed, and these will by dealt with in stratigraphic order, beginning with the oldest known evaporites, from the Archean of the North Pole, Western Australia (Donnelly, Dunlop & Groves, 1978; Dunlop, Groves & Buick, 1979).

NORTH POLE BARITE, WESTERN AUSTRALIA

The North Pole deposits occur in the Early Archean Warrawoona Group. The host rocks to the deposit are considered to be stratigraphically equivalent to the volcanic Differ Formation, which has been dated at 3.45 Gyr, an age which is corroborated by a 3.4 Gyr age for North Pole galena (Richards, Fletcher & Blockley, 1981). The barite occurs in a sequence of chert, arenite and conglomerate which are also silicified, and which are over- and underlain by slightly

Lecture Notes in Earth Sciences, Vol. 13
T.M. Peryt (Ed.), Evaporite Basins
© Springer-Verlag Berlin Heidelberg 1987

metamorphised mafic and ultramafic volcanics. Within the sequence are silicified stromatolites (Walter, Buick & Dunlop, 1980), are some rather controversial microfossils (Dunlop et al. 1978; Awramik, Schopf & Walter, 1983; Buick, 1984). The rocks of the North Pole were first described by Dunlop (1976).

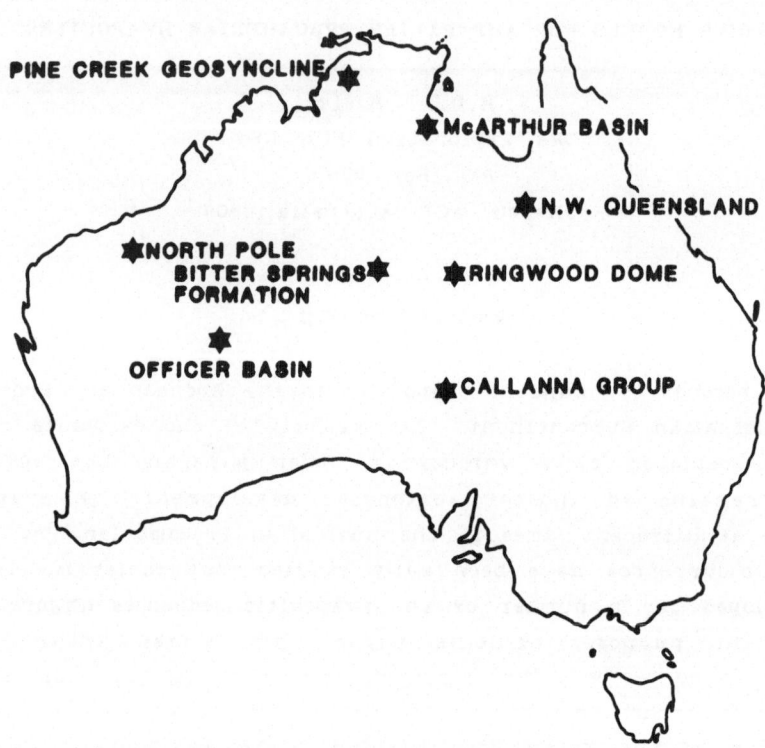

Figure 1. Localities of Australian Precambrian evaporites discussed in this paper.

The barite deposits are extensive, and have been commercially exploited. The beds are bottom-nucleated and may contain crystals 5-20 cm in length radiating across bedding, with overlying laminated chert which drapes over crystal tops, and fills the space between crystals. The top of the crystals in places have been eroded, and clasts of bedded barite occur in intraformational conglomerates, indicating synsedimentary formation of the crystals.

Although the crystals are now barite, Lambert et al. (1978) observed swallow-tail twins, and on measuring the interfacial angles of single non-deformed barite crystals, concluded that the crystal faces were typical of gypsum,not barite. Silica pseudomorphs after single or twinned gypsum crystals are also present, as well as silica pseudomorphs after cubic crystals which may have been halite. Studies

7

of sulfur isotopes also implied that the sulfur was evaporitic,and non-volcanogenic, and formed as a result of surficial oxidation of reduced sulfur from exhalative sources. The authors suggest that the required oxygen was supplied by the action of blue-green algae or sulfur bacteria.

The evidence for an evaporitic gypsum precursor for the North Pole barite is clear. Sedimentary structures in the cherty host rock sequence consist of shallow water to emergent features such as graded and cross-bedding, scouring, intraclasts, ripple marks, and desiccation features such as curved intraclasts, and possible mud-cracks. The chert contains partially replaced carbonate, dolomite rhombs, and rhombic voids. The postulated shallow water to emergent carbonate depositional environment is consistent with an evaporitic origin for the gypsum precursors of the barite crystals.

The facies model envisaged by Lambert et al. (1978) consists of juvenile sulfurous exhalations being added to the basin, followed by microbial oxidation of the reduced sulfur. Bedded sulfate deposits were then formed in shallow water evaporitic environments. This model was subsequently refined by Buick, Dunlop & Groves (1981) (Figure 2), who postulated a closed (barred) basin environment for the deposition of the evaporitic sulfates.

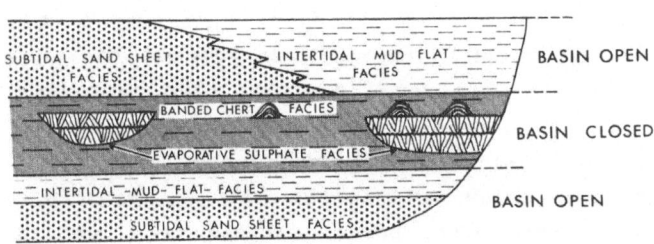

Figure 2. Schematic model for the evolution of the chert-barite unit, North Pole (after Buick, Dunlop & Groves, 1981).

PINE CREEK GEOSYNCLINE, NORTHERN TERRITORY

The rocks of the Pine Creek Geosyncline underwent a metamorphic event at 1.8 Gyr and overlie granitic and gneissic basement complexes dated at 2.47 Gyr (Page, Compston & Needham, 1980). The Pine Creek Geosyncline occurs to the south and east of Darwin, in the Northern Territory (Figure 1), and a summary of the regional geology may be found in Needham, Crick & Stuart-Smith (1980). Coarsely crystalline magnesite and dolomite rocks occur in the Rum Jungle and Alligator Rivers Uranium Fields in the Celia and Coomalie Dolomites at Rum Jungle, and in the Cahill Formation in the Alligator Rivers region. There are > 300 m of these carbonates, which are in places interbedded with pelites and which overlie, at Rum Jungle, a basal arenite unit which is in unconformable contact with the basement.Although the age limits are very broad,the carbonates could be as old as 2.2 Gyr.

The carbonate rocks are abundantly stromatolitic (Crick & Muir, 1979, 1980). Stratiform and domal stromatolites are common, some of the domes reaching a height of 1.5 m, and a similar diameter. Conical stromatolites ("*Conophyton*") are well developed in all there formations, and have a characteristic "organ pipe" appearance on the weathered surface. The stromatolites contain abundant crystal casts which will be discussed shortly, but they are interbedded with non-stromatolitic carbonates which contain cross-bedding, ripple marks, and small scour structures, as well as teepee structures and small intraclasts. The depositional environment indicated by these structures is shallow water to emergent. The stromatolites are interpreted as having developed in supralittoral, littoral or sub-aqueous environments (which are neccessarily shallow too).

The carbonates are coarse-grained because they contain abundant blade-shaped or discoidal crystals, up to 1 cm long. Cruciform and swallow-tail twins are not uncommon. Orthorhombic shapes are also present, and some beds contain abundant cubic casts. The rocks have been described as marbles, implying a metamorphic origin for the crystals. However, if this were the case, then twin lamellae could be expected in the carbonates, and these do not occur. Bone (1983) suggested that the cubic and discoidal shapes were forms of magnesite which crystallised below and above 150° C respectively. However, identical crystal forms have been described from the Amelia Dolomite (Muir, 1979), a formation which has never been subjected to elevated temperature and pressure conditions, certainly not above 100° C. The Amelia Dolomite crystal shapes are interpreted as being after halite (cubic) and gypsum (discoidal), and the Rum Jungle crystals are also so interpreted. Further evidence for the former presence of sulfate at Rum Jungle can be adduced from the presence of lutecite, a variety of length-slow quartz which pseudomorphs anhydrite.

The evaporite minerals must have survived the 1.8 Gyr metamorphic event, since they have retained their primary form. Replacement by carbonate must post-date the metamorphism since there are no signs of strain in the crystals, unlike non-evaporitic carbonates in the area. Anhydrite is known to survive high grade metamorphism (Serduchenko, 1975), and the original gypsum in the Pine Creek rocks would have been converted to anhydrite during burial diagenesis (West, 1964).

A facies model for these deposits was developed by Crick & Muir (1980) (Figure 3).

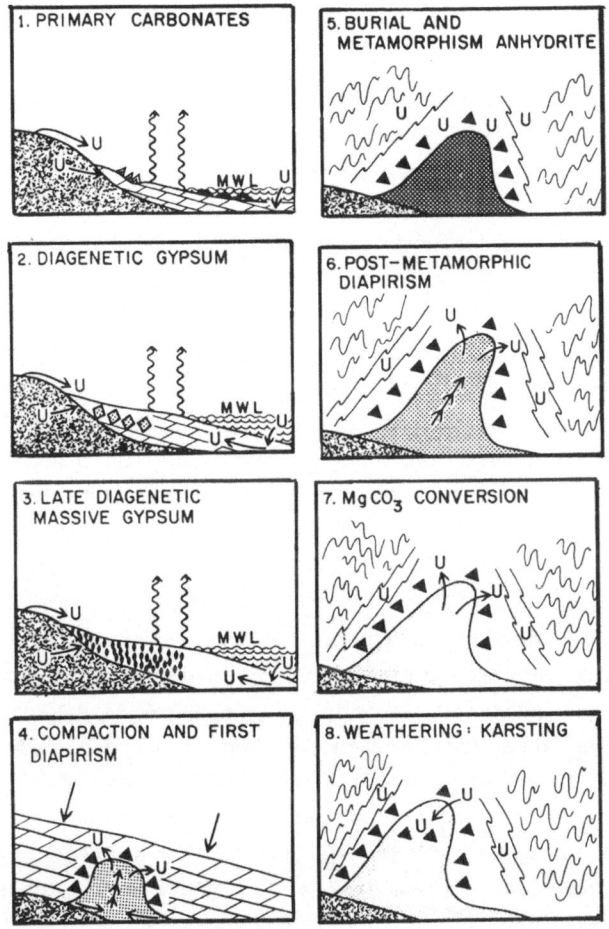

Figure 3. Stages in the development of the uranium deposits in the Alligator Rivers and Rum Jungle Uranium Fields, Pine Creek Geosyncline (after Crick & Muir, 1980).

The crystals habits of both the gypsum and halite indicate that they grew displasively in carbonate muds in shallow water to emergent conditions (from the sedimentary structures already described). Crick & Muir (1980) interpret this environment to have been an arid sabkha (Figures 3.1 and 3.2). Geological evidence in the area indicates that the carbonates have behaved diapirically and the geometry of the diapirs and their relationship to mineralisation suggest that the highly concentrated evaporitic brines may have been a source for the metals in the uranium deposits, and well as in some smaller base metal deposits in the area.

NORTHWEST QUEENSLAND

The rocks of the Corella Formation (1.74-1.54 Gyr, Page, 1983) in northwest Queensland contain abundant crystal casts which have been identified as scapolite. They occur in sequences of black shale, arenite and carbonate rocks which have been metamorphosed to greenschist metamorphic grade.

Because of the metamorphic grade, most of the crystal casts have been identified as "scapolite" and regarded as the products of metasomatic fluids (Ramsay & Davidson,1970). However, recent sedimentological work (Connor, Johnson & Muir, 1982) showed that many of the so-called "scapolite" crystals did not contain that mineral, but hosted a variety of others including quartz, carbonate, sericite, microcline, and undeterminable clay minerals. In addition, there was not one, but many crystal habits displayed by the so-called scapolite. This led Connor et al. (1982) to conclude that the "scapolite" crystals were replacing a variety of original evaporite minerals, most of which had grown displacively in the primary sediment in early diagenesis.

The host sediments contain abundant sedimentary structures, such as cross-bedding, graded-bedding, ripple marks, slump structures and erosional breaks indicated by the presence of fossil caliche deposits. The graded beds were interpreted by Connor et al. (1982) as resulting from turbidity flows in a density stratified lake. None of the graded beds shows any sign of bottom structures or traction currents, as would be expected if turbidity flows ran across the surface of soft sediment. However, if such flows ran across the surface of a denser water body, then at the cessation of flow, the particles would settle through the denser water column producing a perfectly graded bed.

Some of the crystal casts are discoidal and probably pseudomorphs after discoidal gypsum such as would occur in a sabkha environment. In support of this, there are quartz nodules replacing anhydrite, also in the sequence. However, most of the crystal casts appear to have the morphology of shortite, a double carbonate of sodium and calcium which is only precipitated in association with alkaline lake deposits such as the Eocene Green Formation of Wyoming (Eugster, 1980). The presence of shortite pseudomorphs is very significant, because it is impossible for this mineral to precipitate as a result of the evaporation of sea water. It can only precipitate from evaporation of alkaline lake waters. Thus the presence of shortite (or shortite casts) is a proitive

11

indication of non-marine depositional environments.

In the Corella Formation in northwest Queensland, although no detailed facies models have been proposed, Connor et al. (1982) suggested, on the basis of the evaporite casts and the sedimentary structures, that the depositional environment of the black shale was subaqueous, with marginal carbonate sabkhas, and fluvio-lacustrine arenites.

MALLAPUNYAH FORMATION AND AMELIA DOLOMITE, MCARTHUR BASIN

The Mallapunyah Formation and Amelia Dolomite occur in the McArthur Basin (Figure 1) in the Northern Territory. Statigraphically, they lie in the lower part of the McArthur Group (Figure 4: from Muir,1979), and consist of an arenite sequence overlain by a sequence of carbonates and interbedded shale with subordinate arenites.

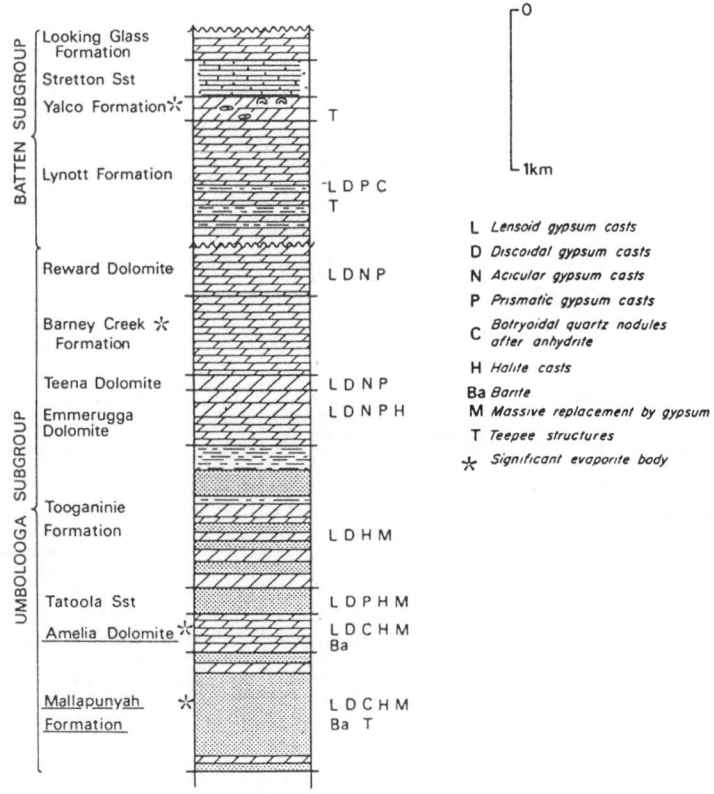

Figure 4. Schematic stratigraphic section through the McArthur Group showing the Mallapunyah Formation, Amelia Dolomite, Barney Creek Formation, and Yalco Formation, all of which contain evaporite deposits.

The presence of abundant relic evaporites in the McArthur Group was first noted by Walker et al. (1977), and a sabkha model for the Amelia Dolomite and Mallapunyah Formation was proposed by Muir (1979). A precise age for the deposit has not been determined, but it is older than the 1.68 Gyr age determined for the H.Y.C Deposit (Page, 1981).

Halite casts are very common in all formations of the McArthur Group, and particularly so in the Amelia Dolomite and the underlying Mallapunyah Formation. Many of the halite casts occur on the bottom surfaces of arenites and formed after dissolution of halite precipitated as a result of evaporation of saline surface waters. Other casts, however, are chert or carbonate replacements of halite cubes that grew interstitially in fine carbonate mud in early diagenesis (Neev & Emery, 1967).

More abundant by far than the halite casts, are pseudomorphs after gypsum, some showing crystalline form, some displacive discoids. All have flakes of the original sulfate in them, but now the crystals consist of dolomite or magnesite, and are variably ferroan. The crystal habits are consistent with deposition in a sabkha environment. This interpretation is reinforced by the abundant occurrence of botryoidal quartz nodules, a replacement of nodular anhydrite which is typical of the arid sabhka of Abu Dhabi in the Arabian Gulf (Muir, 1979).

| Height in section (m) | Lithology | Environment | | Lithology | Environment | | |
		Primary	Diagenetic (early)		Primary	Diagenetic (early)	Vadose and late diagenetic
Lagoon waters	Algal mat, primary gyp. arag. calc.	Subtidal to intertidal	Intertidal				
Sabkha surface							
13	Algal mat, gyp. cement. Qtz. sand	Supratidal sabkha	Inner flood recharge zone	Stratiform stromatolites Beach/splash travertine	Supratidal to high supratidal.	Dol./chert small gyp pseudo-morphs	Dol. recryst/ recemented
12	Arag./cal. muds gypsum mush on		Intermed. flood recharge zone				
11	surface. gy. cemt. Algal mat + gyp cemt. anh.		Outer flood recharge zone				
10	nodls.						
HOLOCENE							AMELIA DOLOMITE
9							
8	Unconsolidated, uncemented aeolianites. Large anh. nodls. 3 x 1 m. Gyp. interstitial cemt. Halite, sylvite, poly- halite, and detrital clay in lenses and beds	Terrestrial	High supratidal zone	Dol. shale/silt ?eolian Cauliflower chert, cubic casts, K-feldspar beds	Terrestrial/ shallow	Gyp./anh. halite, anh. nodls. ?polyhalite, ?sylvite, high supratidal Dol./chert	Dol. recryst/ recemented High K in clay- stones gives K-feldspar-anh. nodls. silicified
7							
6							
5							
4							
3							
2							
1							
PLEISTOCENE							MALLAPUNYAH FORMATION

Table 1. Stratigraphy of the Abu Dhabi sabkha (after Butler, 1969), compared with that of the Amelia Dolomite and Mallapunyah Formation.

A comparison of the uppermost part of the Mallapunyah Formation and lowermost Amelia Dolomite with the Pleistocene and Holocene of the Arabian Gulf (Butler, 1969) shows striking similarities (Muir, 1971; Table 1). The lithologies for both are similar, and the interpreted primary and diagenetic environments are the same. Since this comparision was first published, Patterson & Kinsman (1981) have made a new interpretation of the hydrology of the Abu Dhabi sabkha. In earlier studies, it was assumed that the source of the sulfate for the anhydrite was sea water, but the new hydrological studies proved that it was, in fact, continental groundwater. Thus the identification of a fossil analogue to the Abu Dhabi sabkha does not, of itself indicate marine marginal conditions, but could just as readily represent the sediments of a supralittoral lacustrine environment. Thus the conclusion that the basal Amelia Dolomite is shallow water marine (Muir, 1979) may have to be altered to encompass a shallow water lake environment.

H.Y.C. DEPOSIT, McARTHUR BASIN

An account of the geology of the giant H.Y.C. lead-zinc deposit of the McArthur Basin is given in Walker, Logan & Binnekamp (1978). The deposit has been dated at 1.68 Gyr (Page, 1981). For further information on the mineralisation, the reader is referred to papers by Williams (1978 a and b), and Rye & Williams (1981). Williams and Logan (1981) and Muir (1983) have discussed the depositional environment of the deposit in some detail.

The deposit consists of extremely fine-grained lead-zinc sulfide hosted by carbonaceous dolomitic siltstone of the H.Y.C Pyritic Shale interbedded with fault talus beccia of the Cooley Dolomite Member. Early interpretations of the depositional environment were that the sequence was deep water, but the results reported by Williams & Logan (1981) neccessitate a shallow water to emergent depositional environment. These authors based their arguments on the presence of certain features that taken together indicate deposition under shallow lacustrine conditions. The features are (1) lithified crusts, teepee structures, and intraclast conglomerates; (2) carbonate pseudomorphs after nodular anhydrite; and (3) pedogenic pisoids. The combination of lithified crusts, teepee structures and intraclast conglomerates has been described by von der Borch & Lock (1979) from ephemeral alkaline lakes of the Coorong region, South Australia, and from the Yalco Formation by Muir, Lock & von der Broch (1980) (see also this paper). The nodular anhydrite pseudomorphs in the Barney Creek Formation are identical with those described from the Mallapunyah Formation, and are the products of deposition in a sabkha environment. The presence of replaced anhydrite nodules in the Barney Creek Formation indicates that the sediment surface was exposed to the atmosphere, and that concentrated groundwater brines were moving through the sediment in earliest diagenesis. The same conclusion can be drawn from the occurrence of the lithified crusts. The pedogenic pisoids described by Williams & Logan (1981) formed from carbonate deposition in the capillary fringe after emergence of the sediment pile.

A cartoon of Williams' & Logan's (1981) facies model in shown as Figure 5. It is of particular interest because the anhydrite nodules indicate arid sabkha conditions, but the lithified crusts, by analogy with the Coorong ephemeral lakes indicate an evaporitic but seasonally humid environment. Thus the evaporite facies appears to be recording quite subtle climatic variations.

YALCO FORMATION, MCARTHUR BASIN

The Yalco Formation of the McArthur Basin has been described in detail by Muir, Lock & von der Borch (1980). It lies stratigraphically higher than the Barney Creek Formation, and consists of cherty carbonates with a great variety of sedimentary features. These include intraclast conglomerates , ripple marks, crossbedding, mud cracks, stratiform, domal and laterally-linked stromatolites, and oncolites. There are no distinct evaporite casts in the sequence. The chert occurs as nodules or laminae and stromatolites tend to be preferentially silicified.

The Yalco Formation sequence is compared with the Holocene dolomite forming in ephemeral lakes at the southern end of the Coorong Lagoon, South Australia (Muir et al., 1980). On the basis of the sedimentary structures alone, the Proterozoic and Holocene sequences are identical and similar hydrological conclusions can be drawn for both. The Coorong lake system is an evaporitic system with a slow oceanward migration of alkaline groundwater cropping out sequentially in ephemeral lakes whose composition is controlled by their positions in the groundwater regime. The least soluble salts precipitate in the proximal lakes, with progressively more soluble salts precipitating in sequence in the more distal lakes. The composition of the salts varies from hydromagnesite, through dolomite, and aragonite to halite. In the dry summer months, the halite precipitates, but it is redissolved in the wet winter months, and flushed out of the system. Thus although the Coorong ephemeral lakes are genuine sites for deposition of a typical evaporite mineral (halite), the seasonally humid climatic regime

15

ensures that no preservable deposits of halite remain. The carbonates, however, with their particularly distinctive sedimentary structures are preserved, and it is these structures that are so characteristic of the Yalco Formation.

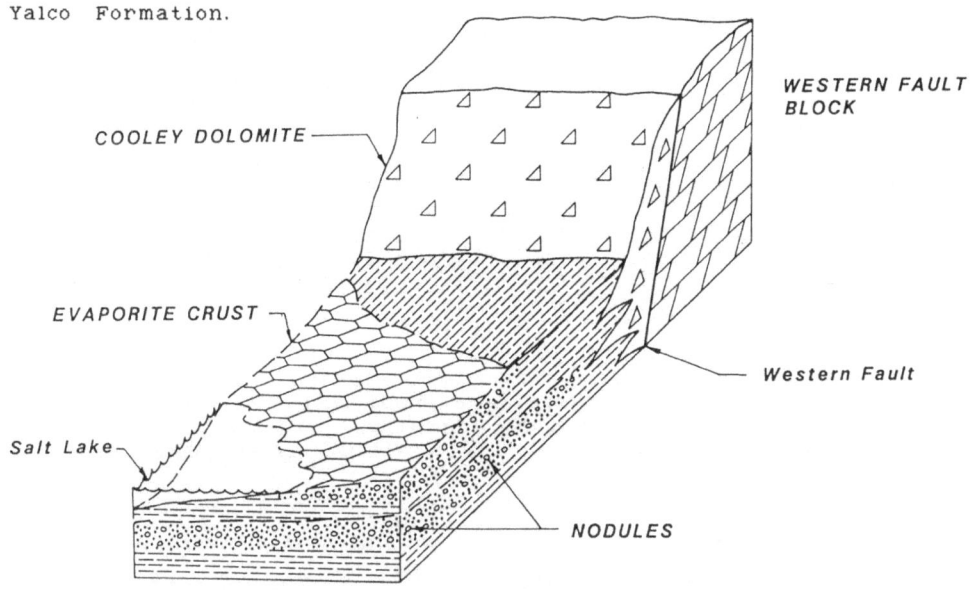

Figure 5. Sedimentation model for the nodular facies, H.Y.C. Deposit (after Williams & Logan, 1981).

The Yalco Formation is part of the Batten Sub-group of the McArthur Group and can be related to an overall facies model (Figure 6) developed for the whole Sub-group (Muir et al. 1980). The formations in the Sub-group range from lacustrine, subaqueous in the Lynott Formation, through littoral to supralittoral and sabkha environments in the upper parts of the Lynott Formation. The ephemeral lake cycles of the Yalco Formation follow, and are themselves succeeded by the fluvial or playa lake fine arenite of the Stretton Sandstone.

CALLANNA GROUP, SOUTH AUSTRALIA

The sediments of the Callanna Group of the Willouran Ranges, South Australia, are imprecisely dated, but are usually included in the later Proterozoic Adelaidean.They are generally accepted as being between 1.4-0.8 Gyr, although recent unpublished results suggest that the younger age may be more accurate. The Group as a whole is described by Murrell (1978) and Preiss, Rutland & Murrell (1981).

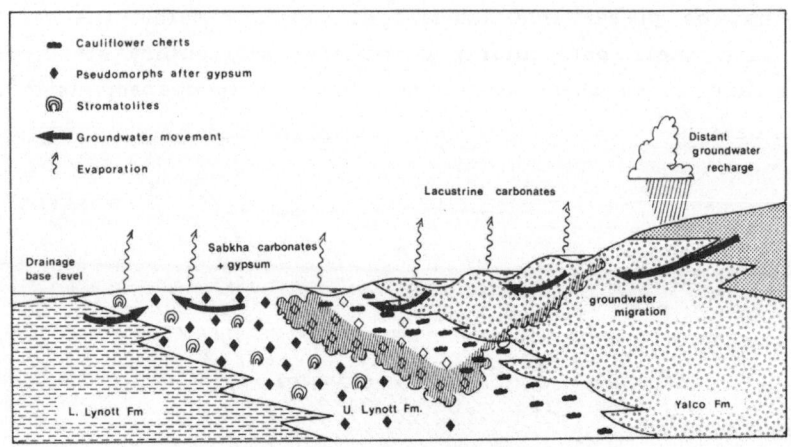

Figure 6. Model for regressive development of the Yalco and contiguous formations of the McArthur Group; note proposed reflux plume (hachured) which invades the upper Lynott Formation, causing leaching with consequent silicification and dolomitization (after Muir, Lock & von der Borch, 1980).

The sequence is characterised by numerous crystal casts which have been described by Rowlands, Blight, Jarvis & von der Borch (1980). The crystal forms that they have described consist of matchstick-like forms up to 10 cm long and 0.5 cm wide. There are in addition, short stumpy and irregular crystals now replaced by microcline. These all have been identified by Rowlands at al. (1980) as pseudomorphs of various crystal forms of shortite, the double carbonate of sodium and calcium that typically occurs as displacive crystals in alkaline playa lake systems. The microcline is believed to be authigenic and having formed in an evaporitic environment. Crystal casts with an arrow head morphology are suggested to be gaylussite ($Na_2CO_3 * CaCO_3 * 5H_2O$), a mineral found in the alkaline Lake Magadi of Kenya.

Discoidal casts, after gypsum, but now microcline, also occur in some beds, as well as rosettes and cauliflower chert nodules after anhydrite.

Rowlands et al.(1980) present a facies model for the evaporites of the Callanna Group that envisages development of playa lakes and sabkhas, in a continental graben system. The continental nature of the graben system is confirmed by the presence of shortite relics, which as stated before cannot precipitate from sea water during evaporation. They compare the Callanna Group rift systems with the modern East African Rift system and point out that alkaline volcanism is common in the Callanna Group, strengthening the comparison.

BITTER SPRINGS FORMATION, NORTHERN TERRITORY

The Bitter Springs Formation occurs in the Amadeus Basin in the southern part of the Northern Territory. It is dated at about 0.9 Gyr (Stewart,1979) and consists of carbonate, chert, and shale with rare volcanics. Large volumes of gypsum and halite occur in the Bitter Springs Formation (Wells, Forman & Ranford, 1965) and in equivalent rocks in in the Officer Basin, Western Australia (Jackson & van de Graaff, 1981) where they have moved diapirically (Browne, Woolnough, and Madley diapirs). However, with the exception of the work of Stewart (1979) on the Ringwood Dome, no facies models have yet been developed for these deposits.

Unlike all the deposits so far described, the Ringwood evaporites are still preserved as gypsum. The system is bipartite with a lower 127 m characterised by brecciated pyritic bituminous dolomite, which contains dolomite gypsum breccia, coarsely crystalline anhydrite, satin-spar gypsum, and chloritic dololutite. The upper part of the evaporite consists of dolomite gypsum breccia, and this is overlain by limestone breccia and massive stromatolitic limestone, which Stewart (1979) interprets as a reef.The gypsum is secondary after anhydrite.

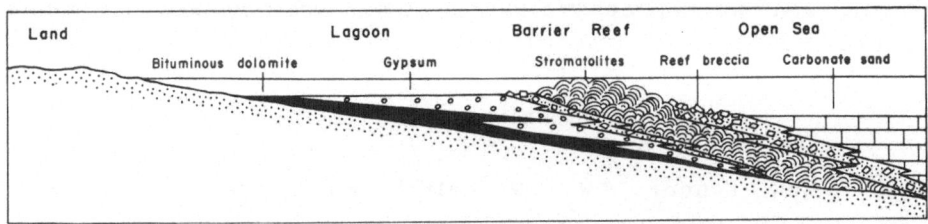

Figure 7. Diagrammatic representation of hypothetical barred basin, showing major rock-types of Ringwood evaporite in lagoon behind stromatolitic reef in Bitter Springs area (after Stewart,1979).

Stewart (1979) develops a barred basin facies model for the Ringwood evaporite (Figure 7). The stromatolite reef cut off a back reef basin from marine waters and the trapped sea water then evaporated precipitating first carbonates and then sulfates, mainly gypsum. The barrier reef then migrated across the evaporite lagoon and prevented the evaporites from being dissolved when normal marine waters transgressed across the lagoon. The primary gypsum was converted to anhydrite, which was later reconverted to gypsum as a result of post-tectonic erosion.

SUMMARY

The facies models presented here represent a range of environments from continental alkaline playas,continental sabkhas,to barred basins. The classic Phanerozoic evaporite styles of the Permian of the North Sea, of the Messinian of the Mediterranean which appear to represent desiccation of deep marine basins have not been recognised in the Australian Precambrian. Not are there any records of potash salts,although the microcline often associated with Proterozoic evaporites may represent the resting place of the potassium from original potash minerals.

ACKNOWLEGDEMENTS

I acknowledge with thanks permission to publish from CRA Exploration Pty. Ltd.; and am grateful to the following individuals and organisations for permission to reproduce diagrams: R. Buick, I.H. Crick, J.S.R. Dunlop, D.I. Groves, D. Lock, C.C. von der Borch, A.J. Stewart, N. Williams, R.G. Logan, International Atomic Agency Commission, International Association of Sedimentologists, Society of Economic Paleontologists and Mineralogists, Geological Society of Australia, and Bureau of Mineral Resources, Geology and Geophysics, Canberra.

REFERENCES

Awramik S.M., Schopf J.W. & Walter M.R., 1982 - Filamentous microfossils from the Archean of Westen Australia. Precambrian Research, 20, 357-374.

Bone Y., 1983 - Interpretation of magnesite at Rum Jungle, N.T.,using fluid inclusions. Journal Geological Society of Australia, 30, 375-381.

Buick R., 1984 - Carbonaceous filaments from North Pole, Western Australia: are they fossil bacteria in Archaean stromatolites ? Precambrian Research, 24, 157-172.

Buick R., Dunlop J.S.R. & Groves D.I., 1981 - Stromatolite recognition in ancient rocks; an appraisal of irregularly laminated structures in an Early Archaean chert-barite unit from North Pole, Western Australia. Alcheringa, 5, 161-181.

Butler G.P., 1969 - Modern evaporite deposition and geochemistry of co-existing brines, the sabkha, Trucial Coast, Arabian Gulf. Journal Sedimentary Petrology, 39, 70-89.

Connor A.G., Johnson I.R. & Muir M.D., 1982 - The Dugald River Zinc-Lead Deposit, Northwest Queensland, Australia. Australasian Institute of Mining and Metallurgy, Proceedings, 283, 1-19.

Crick I.H. & Muir M.D., 1979 - Evaporites and uranium mineralisation in the Pine Creek Geosyncline. International Uranium Symposium on the Pine Creek Geosyncline (Sydney), Abstracts 30-33.

Crick I.H. & Muir M.D., 1980 - Evaporites and uranium mineralisation in the Pine Creek Geosyncline. Proceedings of the International Uranium Symposium on the Pine Creek Geosyncline. International Atomic Energy Agency, Vienna, 1980. 531-542.

Dunlop J.S.R., 1976 - The geology and mineralisation of part of the North Pole Barite deposits, Pilbara region, Western Australia. Unpublished B.Sc. Honours Thesis, University of Western Australia.

Dunlop J.S.R., Groves D.I. & Buick R., 1979 - Evidence for Archaean evaporites. Open Earth, 6.

Dunlop J.S.R., Muir M.D., Milne V.A. & Groves D.I., 1978 - A new microfossil assemblage from the Archaean of Western Australia. Nature, 274, 676-678.

Eugster H.P., 1980 - Geochemistry of evaporitic lacustrine deposits. Annual Reviews of Earth and Planetary Science Letters 1980, 35-63.

Jackson M.J. & van de Graaff W.J.E., 1981 - Geology of the Officer Basin, Western Australia. Bureau of Mineral Resources, Australia, Bulletin 206.

Lambert I.B.,Donnelly T.H.,Dunlop J.S.R. & Groves D.I., 1978 - Stable isotope compositions of early Archaean sulphate deposits of probable evaporitic and volcanogenic origins. Nature, 276, 808-811.

Muir M.D., 1979 - A sabkha model for deposition of the Proterozoic McArthur Group of the Northern Territory and implications for mineralisation. Bureau of Mineral Resources Journal for Geology and Geophysics, 4, 149-162.

Muir M.D., 1983 - Depositional environments of host rocks to northern Australian lead-zinc deposits, with special reference to McArthur River. In: Sangster D.F. & MacIntyre D., "Sediment-hosted stratiform lead-zinc deposits". Mineralogical Association of Canada Short Course Handbook, 8, 141-174.

Muir M.D., Lock D. & van der Borch C.C., 1980 - The Coorong - Model for penecontemporaneous dolomite formation in the Middle Proterozoic - McArthur Group, Northern Australia. SEPM. Spec. Publ. 28,51-67.

Needham R.S., Crick I.H. & Stuart-Smith P.G., 1980 - Regional Geology of the Pine Creek Geosyncline. Proceedings of the International Symposium of the Pine Creek Geosyncline. International Atomic Energy Agency, Vienna, 1980, 1-22.

Neev D. & Emery K.O.,1967 - The Dead Sea. Bulletin Geological Survey of Israel, 41,1-144.

Page R.W., 1981 - Depositional ages of the stratiform base metal deposits at Mount Isa and McArthur River, Australia,based on U-Pb zircon dating of concordant tuff horizons. Economic Geology, 76, 648-658.

Page R.W., 1983 - Timing of superposed volcanism in the Proterozoic Mount Isa Inlier, Australia. Precambrian Research, 21, 223-245.

Page R.W., Compston W. & Needham R.S., 1980 - Geochronology and evolution of the Late Archaean basement and Proterozoic rocks in the Alligator Rivers Uranium Field, Northern Territory, Australia. Proceedings of the International Uranium Symposium on the Pine Creek Geosyncline. International Atomic Energy Agency, Vienna, 1980, 39-68.

Patterson R.J. & Kinsman D.J.J., 1981 - Hydrologic framework of a sabkh along Arabian Gulf. Bulletin American Association of Petroleum Geologists, 65, 1457-1475.

Ramsay C.R. & Davidson L.R., 1970 - The origin of scapolite in the regionally metamorphosed rocks of Mary Kathleen, Queensland, Australia. Contributions to Mineralogy and Petrology, 25, 41-51.

Richards J.R., Fletcher I.R. & Blockley J.G., 1981 - Pilbara galenas: precise isotopic assay of the oldest Australian leads; model ages and growth curve implications. Mineralium Deposita, 16, 7-30.

Rye D.M. & Williams N., 1981 - Studies of base metal sulfide deposits at McArthur River, Northern Territory, Australia, III. The stable isotope geochemistry of the H.Y.C., Ridge, and Cooley Deposits. Economic Geology, 76, 1-26.

Serduchenko D.P., 1975 - Some Precambrian scapolite-bearing rocks evolved from evaporites. Lithos, 8, 1-7.

Stewart A.J., 1979 - A barred-basin marine evaporite in the Upper Proterozoic of the Amadeus Basin, Central Australia. Sedimentology, 26, 33-62.

von der Borch C.C. & Lock D., 1979 - Geological significance of Coorong dolomite. Sedimentology, 26, 813-824.

Walker R.N., Logan R.G. and Binnekamp J.G., 1978 - Recent geological advances concerning the H.Y.C. and associated deposits, McArthur River. N.T.. Journal Geological Society of Australia, 24, 365-380.

Walker R.N., Muir M.D., Diver W.L., Williams N. & Wilkins N., 1977. Evidence of major sulphate evaporite deposits of the Proterozoic McArthur Group, Northern Territory, Australia. Nature, 265, 526-529.

Walter M.R., Buick R. & Dunlop J.S.R., 1980 - Stromatolites 3.400-3.500 Myr old from the North Pole area Western Australia. Nature, 284, 443-445.

Wells A.T., Forman D.J. & Ranford L.C., 1965 - The geology of the northwestern part of the Amadeus Basin, Northern Territory. Report Bureau of Mineral Resources Geology and Geophysics, Australia, 85.

West I.M., 1964 - Evaporite diagenesis in the Lower Purbeck Beds of Dorest. Proceedings of the Yorkshire Geological Society, 34, 315-330.

Williams N., 1978a - Studies of base metal sulfide deposits at McArthur River, Northern Territory, Australia. I. The Cooley and Ridge Deposits. Economic Geology, 73, 1005-1035.

Williams N., 1978b - Studies of base metal sulfide deposits at McArthur River, Northern Territory, Australia. II. The sulfide-S and organic-C relationships of the concordant deposits and their significance. Economic Geology, 73, 1036-1056.

Williams N. & Logan R.G., 1981 - Geology and evolution of the H.Y.C. stratiform Pb-Zn orebodies, Australia. Abstracts 3, Geological Society of Australia, 5th Australian Geological Convention, Perth, 1981. "Sediments through the Ages". 8.

CHARACTERISTIC AND ENVIRONMENTS OF SINIAN EVAPORITE
IN SOUTHERN SICHUAN, CHINA

Xi   XiaoSong
Chengdu Institute of Geology
and Mineral Resources
Chengdu, Sichuan
China

INTRODUCTION

During prospection for hydrocarbons in the southern part of Sichuan province in the 1970', Sinian rock salt (of thickness up to 200 m), interbedded with algal dolomite, intraclastic dolostone and anhydrite was discovered. This discovery provided a new example for the late Proterozoic evaporites in Asia. The aim of this communication is to present an information to the geologists interested in evaporites elsewhere.

GEOLOGIC-TECTONIC BACKGROUND

The Sinian in South-West China including the southern part of Sichuan is the first cover in Upper Yangtze Platform after Yangtzeian orogenic cycle at the end of late Proterozoic. Early Sinian consists of continental clastic rocks and glacial deposits. Its basal contact with the underlying old land core composed of epimetamorphic rocks and granite is angularly unconformable due to Jinning orogenic movement. Doushtuo Formation of early late Sinian is composed of littoral clastic rocks and phosphorite. Dengying Formation of latest Sinian consists of shallow marine carbonate rocks, in part evaporites, overlain conformably by Cambrian with hyoliths and trilobits.

Lecture Notes in Earth Sciences, Vol. 13
T.M. Peryt (Ed.), Evaporite Basins
© Springer-Verlag Berlin Heidelberg 1987

24

After Jinning orogenic movement, the basal land core in South-West China formed the Upper Yangtze Platform, on the north of which there was the Kunlun-Qinlin Trough. The platform was situated to the east of the Songpan islands, to the north of the Central Yunnan old land, and to the north-west of the Niushoushan old land. To the east of the platform was the Jiangnan submarine elevation which was neighboured with the open-sea shelf.

During late Sinian there were two groups of old faults with strikes of north-east and east-west in the west side of the platform, along the margin of the islands. The area of the faults cross formed a rhomb fault subsidence and the Changning saline basin was right located in its centre filled with dolostone, nodular anhydrite and rock salt about 1000 m thick (Figure 1). Therefore the distribution of sedimentary facies and the formation of the saline basin was controlled by the paleotectonic framework.

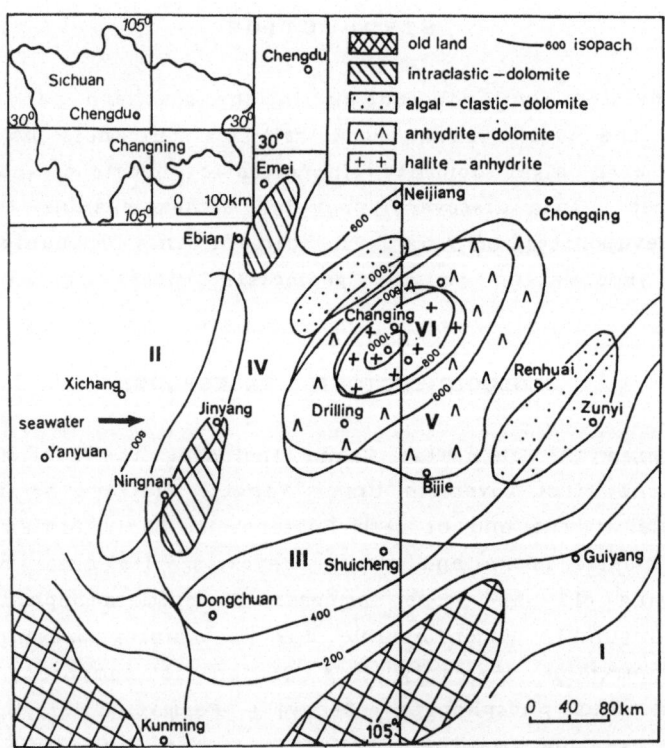

Figure 1. Sedimentary facies in the Dengying Formation of the late Sinian in southern Sichuan.
I - open sea limestone facies, II - intertidal-subtidal dolomite facies, III - intertidal dolomite facies, IV - intertidal algal dolomite facies, V - supratidal sabkha algal dolomite-anhydrite facies, VI - sabkha-salt lake evaporite facies.

ROCKS AND THE SEQUENCE OF THE SALINE SECTION

Dengying Formation consisting of two types of rocks can be divided into three members. The top of the formation (the third member) was subjected to paleoerosion, leaving a residue some dozens of metres thick. The second member called the rich-algal layer is 453-686 m thick and consists of algal dolostone and stromatolitic dolostone, in part possibly evaporites. The first member is 100-500 m thick and is composed of dolomitic limestone, dolostone, algal-detrital dolostone and evaporites (Figure 2), including rock salt that in Changning is 46-200 m thick.

CARBONATE ROCKS

There are four subtypes by composition, structure and origin.
Algal - stromatolitic dolostone. It includes algal-laminated dolostone, algal dolostone, and grape-algal dolostone. These are intertidal sediments and are distributed over the platform, occurring mainly in the second member of Dengying Formation.
Intraclast dolostone. There are mainly two kinds of intraclast dolostone. (a) Sparry dolostone - it consists of algal fragments, oncoids and algal ooids which formed on the high-energy shoals and occurs in the second member of Dengying Formation; (b) Micritic dolostone with different intraclasts and algal pelletoid dolostone with terrigenous sands-they were deposited in subtidal lower-energy zone and intertidal zone, and mainly occur in the first member of the formation.
Supratidal dolostone. It includes bird's-eye dolostone, bamboo-leaves dolostone and nodular, laminated anhydrite dolostone formed in supratidal zone or sabkha. It occurs in the first member.
Others. Dolomitic limestone and black dolomitic argillite formed in the lower energy subtidal environment and occur in the first member of the formation.

EVAPORITES

Halite. Drillholes Ning1 and Ning2 are considered as an example. Halite is stratified and its crystals are 5-7 cm in size. There are pure halite, anhydrite halite and glauberite halite. Average content of NaCl is 93 %, KCl - 0.1 %, Br - 0.0088 %(0.0044-0.001 %); the variation of bromine-chlorine coefficient (Br*1000/Cl) is 0.1-0.15(highest: 0.2) indicating that the halite was precipitated in the early depositional stage of a lagoon or a continental salt lake. The halite crystals are

26

large and few primary sedimentary structures are preserved because of
intense recrystallization and secondary enlargement.

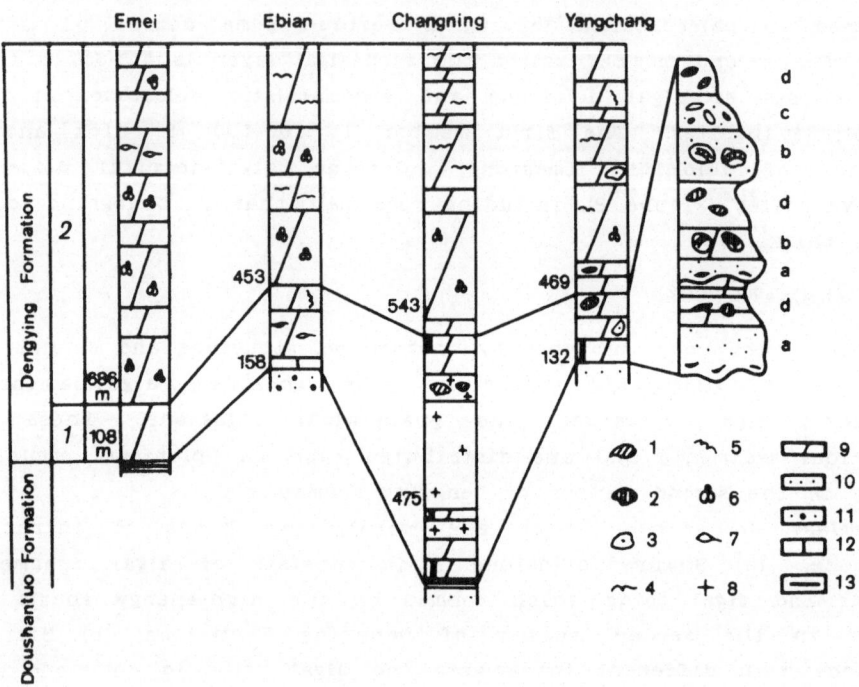

Figure 2. Lithofacies correlation and types of sequence of evaporite in
Dengying Formation.
1 - glauberite, 2 - anhydrite, 3 - oncolite, 4 - algal stromatolite,
5 - algal laminated limestone, 6 - grapestone lump, 7 - bird's-eye,
8 - halite, 9 - dolomite, 10 - quartz sandstone, 11 - sand gravel,
12 - limestone, 13 - siliceous rock, a - sandy algal dolomite,
b - algal dolomite, c - edgewise dolomite, d - nodular anhydrite.

Anhydrite. It mainly occurs in the halitic layers as its
accompaniment. Anhydrite is spotted, cunular-sphaerolic, and lumpy
nodular, forming a series of cycles in halitic layer. The spotted
anhydrite in the lower halitic layer of the cycles is sometimes
extended as a stripe parallel to the stratification. The occurrence of
the anhydrite suggests it formed in supratidal sabkha/lagoon.
Glauberite. Glauberite is concentrated in the interval some dozens
of metres in the top of the halite unit. It is spotted and lumpy. Its
appearance shows that continental fresh water flowed into the salt lake
or the halite in the playa was leached by penetrating surface fresh
water and rainwater at that time, causing the increase of calcium ion

and the interaction between the fresh water and intercrystalline brine to form the glauberite. The formation the glauberite is a mark for the transition from marine to continental deposition. According to the salt-forming processes, the constitution of the saline section in the first member of Dengying Formation can be divided into two types of cycles: (a) Asymmetrical cycle consisting of dolostone-anhydrite-halite, and (b) Symmetrical cycle composed of dolostone-anhydrite-halite-halite with glauberite-dolostone with anhydrite, representing the evolutional sequence of the tidal flat and the salt lake. The section of the first member of the formation lain to the west and south-west of Changning area is matched with the saline strata and can be tripartite: the lower-siliceous dolostone, anhydrite dolostone with quartz sands; the middle-algal-pelletal dolostone with quartz sands, having basal erosion surface and desiccation cracks; the upper-algal potted dolostone, bamboo-leaves dolostone and nodular anhydrite consisting of some dozens of cycles. The tripartite constitution suggests the regressive sequence of evolution of tidal flat/salt flat (sabkha) from the subtidal carbonate environment. The second member of Dengying Formation covering stably all the platform is composed of algal-laminated dolostone, algal-oolitic dolostone, algal grape dolostone, and oncolitic dolostone which formed the high-energy intertidal shoal, showing the transgressive sequence.

ANALYSIS OF THE EVAPORITE DEPOSITIONAL MODEL

*The depositional model of lagoon-salt lake.*The lower salt layers in the first member of the formation are taken as a typical example. There are three types of evaporite sequence (from bottom to top):

(1) Lump anhydritic halite --- cumular-sphaerolic anhydritic halite --- potted anhydritic halite.

(2) Stripped anhydritic halite --- pure halite.

(3) Lump anhydritic halite --- cumular-sphaerolic anhydritic halite --- pure halite

*The depositional model of sabkha-salt lake.*The upper salt layers in the first member of the formation are taken as a typical example. There are two types of evaporite sequence:

(4) Pure halite --- stripped glauberitic halite.

(5) Halite with anhydrite --- potted glauberitic halite.

The five types of evaporite depositional sequence and two types of the saline strata mentioned above show that the evaporite deposition took place in the carbonate platform with dry climate. The upwarping of the upthrow side of the old faults formed the salt flat (sabkha). The subsidence of the cross area of the faults formed lagoon, followed by the playa and the continental salt lake.

The evaporite in the platform mainly deposited on the opposite side of the seawater supply direction. During the Doushantuo stage of late Sinian, the Upper Yangtze Platform was intruded by seawater from the south-west of Yanyuan county. There was gypsum depositing in the part of Ningnan and Jinyang. The transgression beginning with the Dengying stage would make the gypsum migration toward the east. The late regression resulted in the formation of the gypsum sabkha in the part between Yanjin and Luzhou. The whole platform zoning is of tear-drop pattern. But halite deposited in the gypsum flat shows a bull's-eye type of salt lake or playa (Figure 1), which extends north-eastward in the same direction as the strike of the old faults.

PALEOGEOGRAPHICAL ENVIRONMENT

The characteristics of the saline strata, the sedimentary facies sequence of evaporite and the rock structure suggest that the southern part of Sichuan was a shallow carbonate platform, with the islands around which have been leveled down during the deposition of the first and second member of Dengying Formation, so that the terrigenous sands are seldom and most of the platform was influenced by tides. At that time, both the east and west sides of platform were joined with the open sea shelf, the sea water being supplied from the west side. During the period of the regression, the platform was exposed. The Jiangnan submarine elevation to the east of the platform and the marginal shoals of the platform restricted a possibility of seawater to enter. Abundant algal dolostone and other dolostone structures indicate an environment of supratidal algal flat, in which the dolomitization took place. Later, algal growing was prevented because of the intensive evaporation with the dry climate, causing the transition from dolomite to evaporite flat with a lot of nodular anhydrite. In the relatively depression there was the shallow salt lake in which huge thickness of halite originated because of syndepositional block fault subsidence. During the late stage of the evaporite deposition, the marginal shoal of the salt lake became the highland due to the short regression, which stopped the seawater supplying, so that the salt lake in the supratidal sabkha has evolved into the continental salt lake.

The occurrence of the glauberite also proves that the evolution of the salt deposition ended in the continental lake. The thin-banded glauberite formed by the intermixing of residual brines with continental fresh water, but the potted and lump glauberite was formed by replacing the dolostone under the interaction between the fresh water penetrated and intercrystalline brine during the early diagenetic stage after deposition of halite. The occurrence of algal interbedded dolostone in the lower halite layer, the ratio of thickness between the halite and anhydrite (24:1), and the content of bromine in the halite all indicate that the halite in Changning area was precipitated directly from the seawater.

The area depositing halite is less than forty thousand sq. km. The salt basin was small, the brine supply was not sufficient, the process of evaporation was short, and the concentration of the brine was not so high. These resulted in that the deposition of halite was unstable and that thickness of halite layers varies considerably. During the late Dengying stage, the transgression was expanded and seawater covered all the platform in which algal dolostone was deposited, ending thus the evaporite deposition.

THE LOWER CARBONIFEROUS (VISEAN) EVAPORITES IN NORTHERN FRANCE AND BELGIUM: DEPOSITIONAL, DIAGENETIC AND DEFORMATIONAL GUIDES TO RECONSTRUCT A DISRUPTED EVAPORITIC BASIN

J.M. Rouchy*, A. Laumondais** and E. Groessens***

*U.A. 1209, Laboratoire de Géologie, Muséum National d'Histoire Naturelle,43, rue Buffon, 75005 Paris, France.

** TOTAL C.F.P., B.P. 47, 92069 Paris la Défense, France.

*** Geological Survey of Belgium, 13, rue Jenner, 1040 Brussels, Belgium.

INTRODUCTION

In the Franco-Belgian part of Hercynian orogene, presently isolated thick Dinantian anhydritic formations were discovered in two wells (Fig. 1) : Saint-Ghislain in Belgium and Epinoy 1 in northern France (Dejonghe *et al*, 1976; Delmer, 1977; Groessens *et al*, 1979; Rouchy et al, 1984a/b; Laumondais *et al*, 1984; Rouchy, 1986) ; widespread extended breccia, the "Grande Brèche de Dinant et de Namur", and numerous pseudomorphs of gypsum or anhydrite have been observed in their stratigraphic equivalents in boreholes (Wépion, Douvrain, Heugem) and in outcrops (Bless *et al*, 1980; 1981; Swennen *et al*, 1981; Swennen and Viaene, 1985; Hance et Hennebert, 1980; Hennebert and Hance, 1980; Conil et Groessens, 1986; Groessens *et al* 1979; Rouchy, 1986; Rouchy *et al* 1984 a/b; 1986 a/b). In order to reconstruct the original character of these formations and to understand their tectonic impact, a detailed sedimentological and geochemical isotopic study was carried on the three groups of sediments : thick anhydritic formations, scattered pseudomorphs and breccias (Pierre *et al* 1984 ; Pierre, 1986;

Lecture Notes in Earth Sciences, Vol. 13
T.M. Peryt (Ed.), Evaporite Basins
© Springer-Verlag Berlin Heidelberg 1987

Pierre and Rouchy, 1986; Rouchy, 1986; Rouchy *et al*, 1984, 1986b) . This study reveals that the present distribution of evaporites is controlled (with local variations) by post-depositional parameters such as tectonism and dissolution, dissecting a regionally widespread unit, which extended in all the structural unit of this part of the Hercynian orogene

STRUCTURAL AND STRATIGRAPHIC SETTING OF EVAPORITES

The Variscan area in Belgium and northern France may be broadly divided into three major structural units (Fig. 1) : 1) the Brabant Massif, which is a fragment of the Caledonian orogen ; 2) surrounding this first unit, the autochthonous Variscan area is formed by the Namur Basin in the South and the Campine Basin in the North [in this part, the Devono-Dinantian ends by a thick coal formation (Borinage ; Campine)] ; 3) in the South, the Dinant Nappe known from the Ruhr to the Ireland is carried , in the South-North direction, over the Synclinorium of Namur by a major overthrust, the "Faille du Midi". This nappe may be traced over 125 km to its roots in the northern part of the Paris Basin (Aubouin, 1985; Cazes *et al*, 1985). The regional structural pattern, already well-known through petroleum exploration (C.F.P.(M), CO.PE.SEP., R.A.P., S.N.P.A., 1965) can be treated as thin-skinned tectonics (Laumondais *et al*, 1984; Becq-Giraudon, 1983; Cazes *et al*, 1985). The structure results from polyphased tectonic activity in which the major thrusting phase (Asturian phase) is late-Stephanian (Colbeaux *et al*, 1977 ; Becq-Giraudon, 1983). Below the main overthrust and in front of it, a multiple-system of thrust slices is observed in paraautochthonous and autochthonous series, particularly in the coal formation. In the studied area, a broad period of denudation occurs during the Permian and partly the Triassic. The Mesozoic Paris Basin overlies the southeastern part of the Hercynian area and a small Cretaceous basin is developed in the central part of the Synclinorium of Namur (Mons Basin) including the type section of the Montian stage of the lower Coenozoic.

In the studied area, the evaporites occur at several stratigraphic levels within the

Devono-Dinantian column but the thickest anhydritic formations are found in the Givetian of three wells (Tournai, Vieux-Leuze and Annapes 1) in the northern edge of the Synclinorium of Namur (Coen-Aubert *et al*, 1980; Rouchy, 1986) and in the Dinantian of Saint-Ghislain and Epinoy 1 wells where respectively 765 m and 904 m of limestone and anhydrite deposits were encountered (Fig. 2). Elsewhere, pseudomorphs of gypsum and anhydrite or very thin evaporitic deposits occur in the Givetian (Preat and Rouchy, 1986), the Frasnian and the Famennian (Graulich 1963; Groessens *et al*, 1979; Goemaere *et al*, 1985; etc) and in the above-mentioned Dinantian occurrences. The pseudomorph-rich sediments as well as the main Visean breccia offer a good stratigraphic correlation with the anhydritic levels of Saint-Ghislain; the Middle Visean (V2b-V3a interval) contains the main anhydritic intercalations, the Great Breccia (Grande Brèche de Dinant et de Namur) and numerous evaporite pseudomorphs (Bless *et al*, 1980; Groessens *et al*, 1979; Conil and Groessens, 1986; Rouchy, 1986).

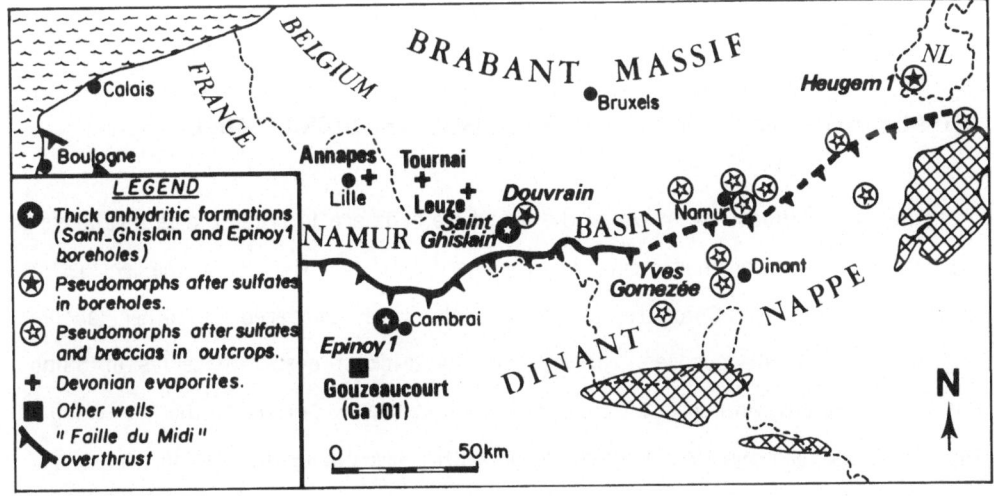

Fig. 1 - Location of wells and localities mentioned in the text

Until recently, the distribution and the significance of these evaporites, particularly the Dinantian evaporites, were underestimated and poorly understood because of limited data in spite of the paper of A. Delmer (1972) who insisted on importance of evaporites for the understanding of the regional geology. The limestone and anhydrite Visean deposits of Saint-Ghislain located in an area of Devono-Dinantian thickening, was first considered as deposited in a narrow subsiding trough (the "Sillon Borain") in front of the Hercynian overthrust. A sedimentological study (Rouchy *et al*, 1984 a/b) concluded that the Saint-Ghislain formation constitutes a residual fragment of a widespread formation extending southward ; this hypothesis was confirmed by the discovery of an another thick anhydritic formation in the Epinoy 1 well below the "Faille du Midi" overthrust, where this formation is situated in inverted position and in a complex slice thrust system (Laumondais *et al*, 1984) ; the evaporitic sedimentation could then have begin in Tournaisian and continued into Visean (Fig. 2). General sedimentological research carried out in Belgium and the Netherlands have revealed the presence of numerous evaporitic pseudomorphs in all the structural units shown in Fig. 1 : autochthonous in the south and the east of the Brabant, Boulonnais, Dinant nappe ; evaporites of the same age are known in Great Britain (Giffard, 1922-23; Georges, 1963; Llewelyn and Stabbins, 1968; Llewelyn *et al*, 1968; West *et al*, 1968).

INTERCALATED LIMESTONES PALEOENVIRONMENTAL SIGNIFICANCE

The Visean limestone in which the anhydrite beds are intercalated is a relatively homogeneous formation characterized by the scarcity or even the lack of the terrigenous components (Fig. 2). Dolomite is surprisingly rare compared to other ancient carbonate/evaporite sequences, at least in the thickest evaporite series of Saint-Ghislain and Epinoy. An important phase of dolomitization is related to the fracturing of limestone has been observed in association with a breccia in the Wépion borehole.

Although the faunal association is poorly diversified, Groessens *et al* (1979) indicate the presence of fossils in all parts of the Saint-Ghislain formation (radiolarians, foraminifers, ostracods, serpulids, bryozoans, crinoids, brachiopods, gasteropods, corals, even goniatites and algaes) and, at a few levels, the fauna appears to be rich. In contrast, some organisms (ostracods and brachiopods) are scarce in the cores of the Epinoy 1 well.

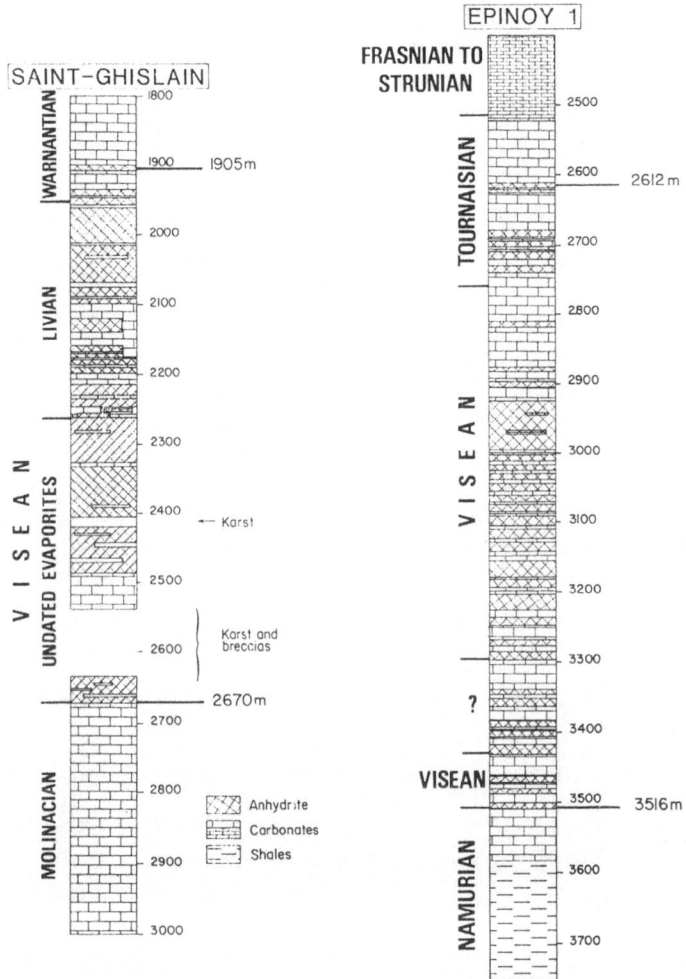

Fig. 2.- Schematic stratigraphic columns of the Saint-Ghislain and Epinoy I boreholes; note the presence of a deep karst at Saint-Ghislain and the reversed position of the Epinoy I series in which we can observe an intensely deformed interval corresponding to the intersection of the thrust slices (between 3,300 and 3,400 m).

Planar or undulating laminated limestones of possible cryptalgal nature are common (Fig. 3, A,B,C). Columnar stromatolites (10 cm in height) have been only observed in one level at Saint- Ghislain (Fig. 3,B). In outcrop, Mamet *et al* (1986) observe the frequent occurrences of the Spongiostromata facies. By the study of the faunal changes in some Visean outcrops, Hance and Hennebert (1980) showed sequences of progressive

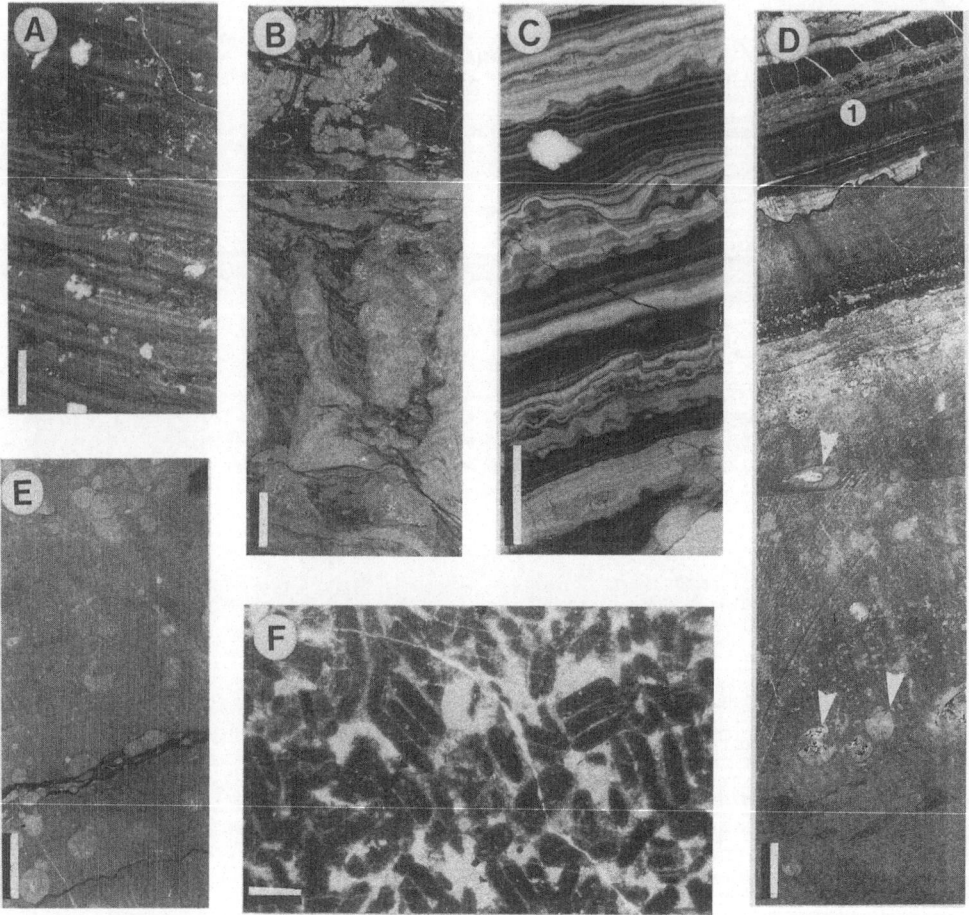

Fig. 3.- Facies of the intercalated carbonates.   A.- Planar cryptalgal laminites; the intraformational brecciation could be related to desiccation. Saint-Ghislain borehole; 1,848 m; scale bar is 2 cm. B.- Columnar stromatolites. Saint-Ghislain borehole; 2,265.20 m; scale bar is 2 cm. C.- Dark, finely laminated limestone probably of algal origin. Some diagenetic   anhydrite nodules may be pseudomorphs after gypsum. Saint-Ghislain borehole; 1,959.90 m; scale bar is 2 cm. D.- Sequence beginning by a goniatites-rich (arrows) sediments grading upward into dark laminated sediments (I) prior to the evaporite deposition. Saint-Ghislain borehole; 1,946.60-1,946.80 m; scale bar is 2 cm. E.- Oncolitic limestone. Saint-Ghislain borehole; 2065.40 m; scale bar is 2 cm. F.- Thin section photomicrograph of peloidal limestone (pelsparite). Epinoy I borehole; 3,133.60 m; scale bar is 100 μm.

increase in restriction from a normal marine environment (foraminifers and ostracods) to evaporitic conditions (algal sediments with pseudomorphs after sulfates). This observation argues for a marine origin of the brines which have generated the evaporitic interbeds. Similar changes in depositional conditions could explain in the Saint-Ghislain borehole (1945,15m), the rapid transition between the sediments containing goniatites,the cryptalgal laminites (Fig. 3D) and the nodular sulfates (Rouchy *et al,* 1984a) ; in other cases, evaporites and carbonates are associated without apparent sequential transition. Ooïds and oncolites are present in thin layers (Fig. 3,E) while beds of peloidal limestones are well developed in Epinoy 1 cores (fig. 3, F). Algal facies are widely represented.

Alternating more or less restricted marine and evaporitic conditions are evidenced by the sedimentary and faunistic changes. As is the case in all evaporitic subaqueous environments, the development of the microbial accretions (stromatolites) following the faunal diversity impoverishment, indicates the strongest restriction stage before the salt saturation (Guelorget et Perthuisot, 1983). In the Visean, there is no evidence of halite (or more soluble salts) with the exception of perhaps a carbonate-sulfate breccia in Saint-Ghislain borehole.

The assumed depositional model shows relationships between carbonates and sulfates in a restricted to repeatedly open lagoonal setting; the evaporitic stage, related to strong reduction of the marine influx or barred conditions, is generally characterized by the lowering of the water level; this produces, in weakly-differentiated paleogeography, the emersion of large areas and the persistance of evaporitic conditions in lows or subsiding areas.

MEANING OF THE PRIMARY AND EARLY DIAGENETIC ANHYDRITE FEATURES

Gypsum, the common expression of the calcium sulfate in the surface conditions, becomes unstable at depth as the consequence of temperature increase and is changed into anhydrite; the transformation produces a textural homogeneization with obliteration of the gypsum crystalline structures and the development of nodular and mosaic anhydrite (Rouchy, 1976; Loucks and Longman, 1982). The range of depth depends on various parameters, principally geothermal gradient, tectonism, seismicity and salinity of the connate waters, but it is commonly estimated to the range of 700 - 1000 meters (Murray, 1964). Gypsum can be also altered into anhydrite in surface syndepositional conditions, and therefore confusing interpretation may arise; Rouchy *et al* (1976), Rouchy (1980), Loucks and Longman (1982) and Shearman (1985) discussed this problem with regard to the Miocene and Cretaceous evaporites.

As in all the deeply buried evaporitic formations, anhydrite is the sulfate main component with a nodular morphgology fashion used here in a broad sense including cm to dm-sized isolated nodules, coalescent nodules forming a mosaic structures (Fig. 4,A), contorted nodular or enterolithic features ; in some cases, the host-sediment appears deformed around the nodules as a consequence of early diagenetic growth or perhaps diagenetic evolution. Numerous pseudomorphs of gypsum indicate its former presence; traces of gypsum were identified at depth (3900 m) in Epinoy 1 well (Bouquillon, 1984; Couilloud and Moine, *in* Rouchy *et al*, 1984b). Pseudomorphs of gypsum crystals (Fig. 4,B,C) exhibit two types of habit : isolated lenticular or lozenge-shaped crystals (mm to cm in size) and vertically standing gypsum crystals beds (cm in height) of selenite type (single or twinned crystals); the first characterizes an early diagenetic growth in carbonate mud or in algal laminites (Saint-Ghislain, 1932 m) and the second results from primary subaqueous crystallization on the floor of the depositional basin. Some samples show the transitional stage between a well preserved crystalline habit and rough pseudomorphs or even nodular fabrics. Many nodules with angular forms evidently had gypsum precursors.

The original shape of the crystals is best preserved when the crystals are isolated within a carbonate matrix or when they are replaced by calcite or silica, suggesting the destruction of the sedimentary structure is broadly post-depositional and probably burial-related. As it is evidencied by some samples of Saint Ghislain, the disappearance of the crystal shape and the development of the nodular aspect could indicate that the original gypsum bed was thick and massive.

Considering all data previously discussed, we can demonstrate that the gypsum could has been an important component in the depositional sulfate phase despite the predominant nodular pattern of the anhydrite. These data do not exclude the early diagenetic growth of nodular anhydrite in temporarily emerged carbonate blanket, but this mechanism cannot explain the formation of the whole anhydrite. We shall see the tectonic deformation greatly contributes to destroy the primary structures.

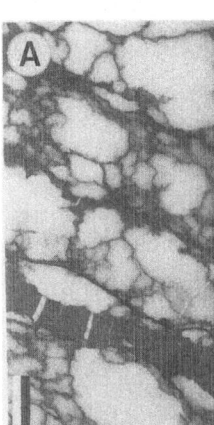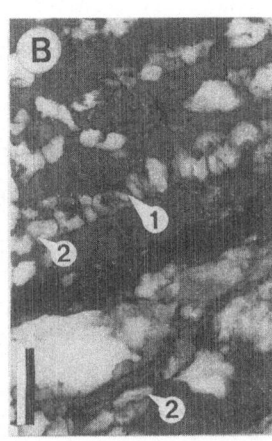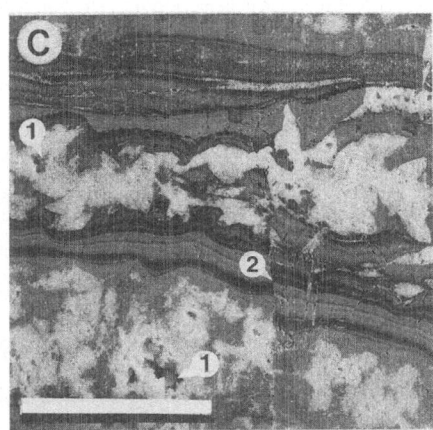

Fig. 4.- Facies and petrography of the sulfates. A.- Typical nodular to mosaïc ("chicken-wire") anhydrite. Saint-Ghislain borehole, 2,219.13 m. B.- Anhydritic pseudomorphs after crystalline gypsum (1); note the transformation of gypsum aggregates erase the primary structure of the gypsum and lead to a "pseudo-nodular" (2) or "pseudo-mosaïc" structure. Saint-Ghislain borehole; 2,108 m. C.- Sparry calcite replacement after lenticular gypsum aggregates which have resulted from early diagenetic growth into black laminated sediments; the voids (1) and perhaps the fracturation (2) probably resulted from gypsum to calcite volume reduction. Saint-Ghislain borehole; 1,932.86 m; scale bar in all photos is 2 cm.

ANHYDRITE TECTONIC DEFORMATION

In the lower part of the Saint-Ghislain formation and in the whole Epinoy 1 series a great variety of deformational structures like stretching, lamination, isoclinal microfolding, augen-like and mylonitic structures (Rouchy *et al*, 1984a/b ; Rouchy,

1986) are generated by compressive tectonic stresses. The similarities between tectonic-generated structures and sedimentary (lamination) or diagenetic (pseudo-nodules) features could lead to incorrect interpretations. Similar structures were also observed in Permian Bellerophon formation of the Italian Alps (Helman and Schreiber, 1985) as in Tuscany (Ciarapica and Passeri, 1976 ; Schreiber and Fitzgerald, oral. comm.) and also in relation with salt migration in many diapiric structures (Wall *et al*, 1961 ; Schwerdtner, 1966 ; Martinez, 1974 ...).

The Saint-Ghislain anhydritic formation shows a progressive downward deformation, in which changes are interpretated as the result of the increasing tectonic stresses. Textural change begins with the rotational deformation of the nodules which become oblique (15-25 degrees) with respect to the layering of the undeformed carbonates (Fig 5, A,B,C); the layers composed of elongated nodules may be separated one from the other by glide planes showing fold offset. The increase of the deformation leads to the tectonic lamination which can be very regular when any carbonate fragment disturbs it (Fig. 5,D); a confusion with similar sedimentary lamination could lead to a misinterpretation. The microscopic observation (Fig. 6, A,B,C,D) reveals the crystal

Fig. 5. - Tectonic and halokinetic deformation of the anhydrite. A. - Stretching and oblique reorientation of the anhydrite layers, intercalated between weakly deformed limestone laminae. Saint-Ghislain borehole; 2,192.40m. B. - Stretched and deformed mosaic structure. Saint-Ghislain borehole; 2,169 m. C. - Stretched mosaic structures between two carbonate beds showing white calcite filled fractures. Saint-Ghislain ; 2,168.46 m. D. - Tectonic lamination the dark laminae are composed of reoriented carbonate fragments (mylonitic structure, 1) resulting of the boudinage of former layers , note the microfolding (2) of some laminae and the presence of isolated isoclinal hinge (3). Epinoy 1 borehole; 2,939.10 m. E. - Augen-like structure resulting from the rotation of dolomitic fragments during the anhydritic flow ; note the drag fold in the anhydrite. Epinoy 1 borehole; 2,939.40 m. F. - Large fractured limestone fragments dragged in the laminated anhydrite ; the penetration of the anhydrite in the open fractures (1) is followed by the boudinage of the thin limestone beds (2). Saint-Ghislain borehole; 2,308.96 m. G. - Fragmentation and microfolding of the calcareous layers. Epinoy 1 borehole; 2,928.80 m. H. - Laminated and microfolded anhydrite. The lamination is deformed by asymetrical chevron folds with slight inclined axis. A dolomitic layer is involved in a isoclinal microfold with axial plane parallel to the lamination. Epinoy 1 borehole; 2,930.70m to 2,931.37 m. Scale bar in all photos is 2 cm.

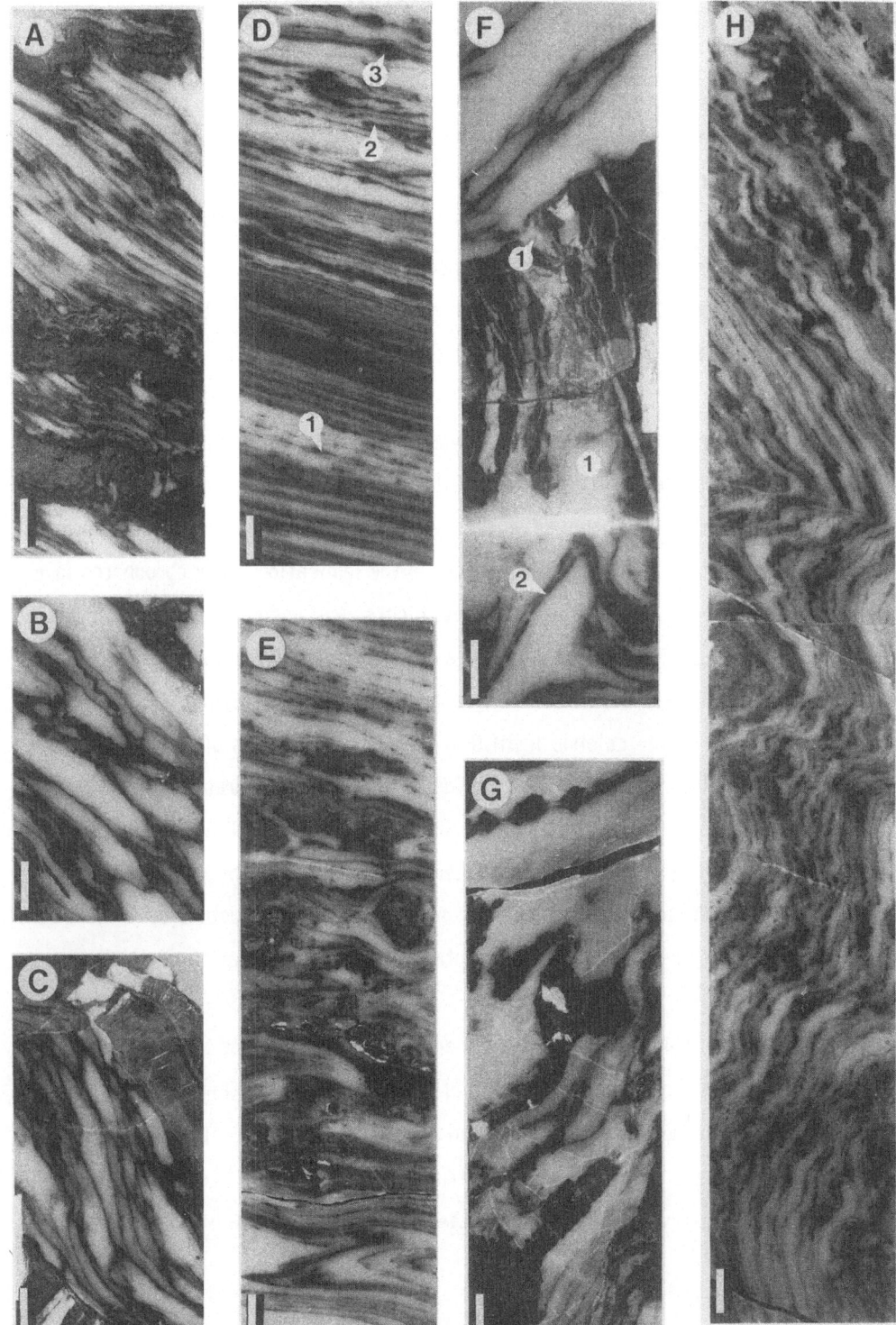

reorientation parallel to the lamination and sometimes, their fragmentation in minute fragments during intragranular gliding and dislocation. Schwerdtner (1974) noted the lineation of the anhydrite rocks in evaporite domes has the same kinematic significance as schistosity in metamorphic tectonites. As this deformational stage, the calcareous interbeds have undergone brittle deformation ; the style and the importance are depending of their relative thickness. The thin layers are stretched and disrupted ("boudinage") or microfolded (Fig. 5,D) ; augen structures (Fig. 5,E) result from the dissemination along the foliation of numerous fractured and rotated fragments of carbonate or of isolated bends of isoclinal microfolds with the axial plane lying parallel to the lamination ; the rotation of some fragments can disturb the regularity of the lamination inducing drag folds in the anhydrite (Fig. 5,E) particularly well developed around competent bodies such as quartz or silicified aggregates, even showing sigmoidal tails of the recrystallized calcite (i.e. pressure shadows). A great number of drawn out hard crystals or fragments produce a disorder in the lamination. Thick carbonates layers are generally brocken and large fractured blocks are enclosed within the flowed anhydrite (Fig. 5,F,G). These blocks present a dense network of fractures (sometimes "en échelons") filled by recrystallized calcite and anhydrite; in Epinoy 1 cores, native sulphur is present in the calcitic infill (Fig. 8,E). Pseudo-nodules of anhydrite seem to result from the injection of flowing anhydrite into open fractures (Fig. 5,F). In more deformed intervals of Epinoy 1, the peloids display oblique stretching. A more confusing fabric which mimics a sedimentary sequence results along the horizontal axis of isoclinal folding in the anhydrite ; indeed, a typical mosaic structure develops in the thickened hinge of the fold and an horizontal and regular lamination characterize the stretched limbs.

Large chevron folds with gently inclined axial planes deforms both the lamination and the isoclinal microfolds near the depth of 2,930 m in Epinoy 1 (Fig. 5, H).

The strongest deformation appears in the lowermost core of Saint-Ghislain in which a very irregular mylonitic-like structure composed of brecciated carbonate fragments associated with flowed and injected anhydrite is seen (Fig. 5,G).

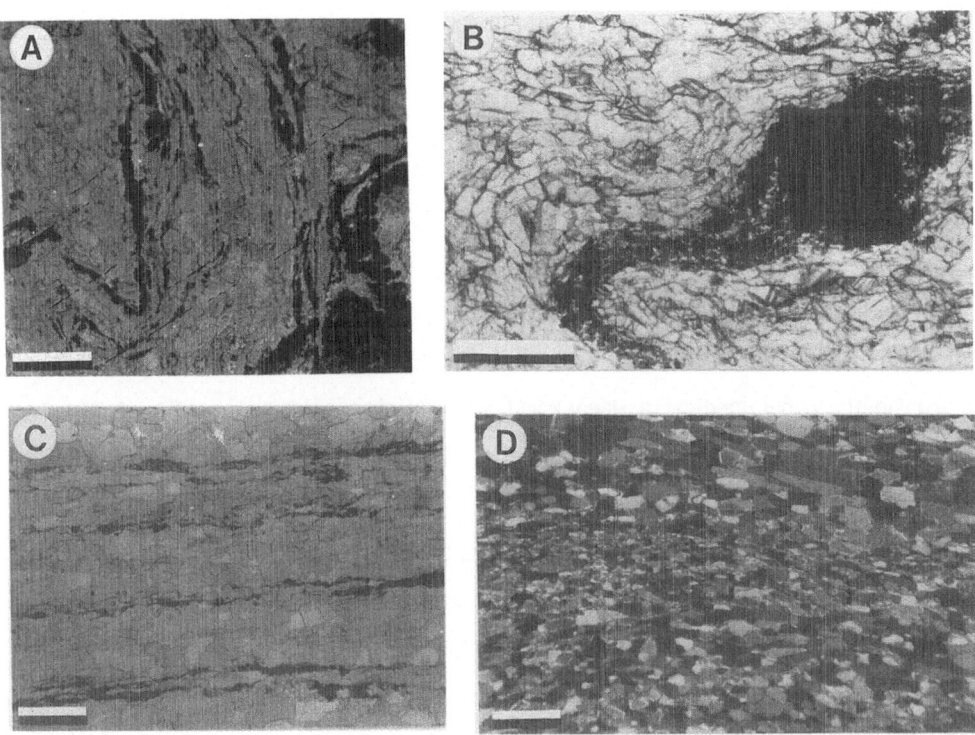

Fig. 6. - Microscopic deformations. A. - Irregularly microfolded anhydrite. Saint-Ghislain borehole ; 2,046,66 m ; plain light. B. - Microfolded anhydrite including a rotated fragment. Epinoy 1 borehole ; 2,928,75 m ; plain light. C. - Regularly laminated anhydrite showing reoriented crystals and stretched (boudinage) limestone laminae. Saint -Ghislain borehole ; 2,048,46 m ; plain light. D. - Laminated anhydrite. Saint-Ghislain borehole ; 2,928,75 m ; crossed nichols ; scale bar in all photos is 500 μm.

The significance of the above mentioned deformation must be discussed with regard to both the structures and to the mechanisms. Firstly, in an inhomogeneous formation composed of an alternation of ductile (anhydrite) and brittle (limestone, accessory dolomite) layers, both the style and the intensity of the deformation vary considerably with respect to the relative thickness of each component; deformation increases with the thickening of the ductile layers. Secondly, the ubiquitous deformational fabrics of the anhydrite cannot allow one to distinguish between regional tectonic compressive stresses and geostatic movements associated or not with former saline beds which produce very similar fabrics (Wall *et al,* 1961 ; Schwerdtner, 1974) ; in their study of the mechanical behaviour of the anhydrite, Müller *et al* (1981) suggest that the deformation of the anhydrite might be yielded by a steady-state flow at relatively low temperature and low stress. We can admit that the tangential tectonism release flow processes as in halokinesis.

Looking the Saint-Ghislain borehole, the downward increase of the deformation to a maximum near the base of the formation suggests it corresponds to a mechanical discontinuity underlined by a deep karst and collapse-solution breccia. In the two boreholes, the regularity of the style of the deformation also with the gentle inclined lamination, except of a part of the Epinoy 1 well seem to be due to transverse differential displacements in relation with tangential tectonic stress (i.e. shear). It is possible therefore, that the structures result from a combination of the Hercynian compressional forces and of the subsequent flow of the evaporite ; the combination of the two mechanisms and the concentration of the forces through the mobility of the evaporites can easely explain the presently observed discontinuity of the evaporitic bodies. The extreme deformational features in the Epinoy 1 formation reflects the more complex structural setting of the formation associated with a multiple-system of slice thrusts below the main overthrust of the "Faille du midi" (Fig. 7). The location of the borehole at the intersection of this main thrust with a SW-NE transverse direction induced by a resistant spur of the Caledonian massif of Brabant (Becq-Giraudon *et al,* 1981 ; Laumondais *et al,* 1984) can have increased the tectonic deformation.

Considering all the data, it can be assumed the evaporites played an active role in the buckling of the regional Hercynian structure of thin-skinned pattern as detachment or gliding layers and more specially in the genesis of duplex structures. Producing mechanical thickening of evaporites in some area and stretching and breaking in others, the associated flow induces the observed discontinuity of the formation (Fig. 7) : in agreement with Delmer's hypothesis (1977), the irregular morphology of domes and troughs of the Paleozoic roof in the Hainaut (basin of Namur) could be due to halokinetic/tectonic flow combined with the subsequent dissolution of the evaporites

(Fig. 7) ; the presence of thick evaporites (Fig. 7) in Saint-Ghislain borehole, situated on the dome edge and of dissolution breccias in Douvrain well recorded in a trough (Leclercq, 1980) confirms this interpretation.

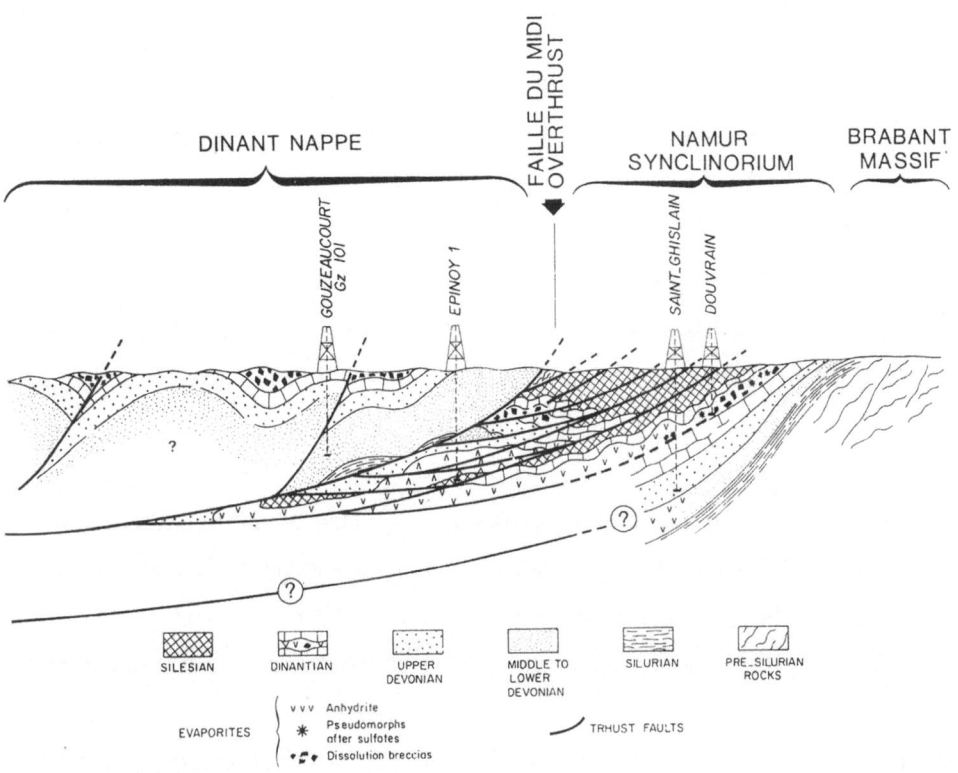

Fig. 7. - Interpretative cross-section showing the possible structural situation of the visean evaporitic bodies.

The example of the Visean anhydrite emphasizes the great importance of the recognition of the anhydrite deformational fabrics as they are similar to sedimentary features and therefore misinterpretations can result.

DIAGENETIC EVOLUTION

Important diagenetic changes observed in Visean evaporitic beds in boreholes as well as in outcrops are influenced by multiple controls during the complex depositional evolution of the series ; among them, we can note: depositional pore-fluid composition, pressure and temperature changes during burial, tectonic deformation, and finally the role of groudwater with the exhumation of the series.

This chapter describes the sequence of crystal authigenesis and diagenetic replacements in order to elucidate the impact of each parameter and to recognize them in the residual series after the tectonic deformation and the dissolution. We have previously discussed the problem of early diagenesis of the calcium sulfate and of the burial-controled gypsum-anhydrite conversion. A more complex set of diagenetic processes leads to the authigenesis of celestite, fluorite and albite, and to the silica and sulfate replacements.

1 -   The accessory mineral authigenesis : celestite, fluorine, albite

Celestite has a wide stratigraphic distribution occurring in Givetian, Tournaisian and Visean in which it is usually present as scattered mm to cm-sized crystals or flabellate crystalline aggregates (Fig. 8,A). The celestite is more abundant in carbonates interbeds than in evaporites where it is nevertheless present in small amounts; in Epinoy 1 cores it forms rare massive centimetric layers of blocky or chert-like microcrystalline aggregates. The inclusion by the prismatic crystals of numerous carbonate relics and even peloids or shell fragments indicates a diagenetic growth in a host-sediment. The celestite crystals can be broken and the layers microfolded during the deformation of the anhydrite.

Among the three kinds of mechanisms generally infered to explain the early diagenetic growth of the celestite: (1) the release of Sr^{++} during aragonite-calcite conversion (Kinsman, 1969) or (2) during gypsum-anhydrite alteration and (3) the direct precipitation, the first and second ones easily explain the diagenetic character of the

Visean celestites.

Fluorite is the next authigenic mineral, specifically associated with Visean strata in the well cores (Saint-Ghislain, Epinoy 1) or in the outcrops (Walhorn, Bomel, etc.). The euhedral to subhedral crystals (70 µm to 400 µm, rarely 1 mm) appear scattered in limestones or in deformed anhydrite (Fig. 8, B), especially in Epinoy 1 core. The corroded boundaries of the crystals in limestones suggest an early diagenetic growth. In deformed anhydrite, the crystals could have been pulled away from fragmented carbonates or could represent relics of a gypsum phase as in some samples of Saint-Ghislain (1848 m) where the fluorite is included in calcitic pseudomorphs after gypsum.

Considering that there is no evidence of hydrothermal activity nor noticeable volcanic contribution in the studied Visean stratas as in some present alkaline lakes with waters of high fluorine content, the genesis of fluorite is classically attributed to two processes : precipitation from saline waters (Sabouraud-Rosset, 1970 ; Sonnenfeld, 1984) or organic pre-concentration (Lowenstam, 1981). The occurrences of fluorite in gypsum crystals or in anhydrite suggest an early diagenetic precipitation with gypsum but we cannot exclude yet another origin.

Albite appears as small euhedral crystals (less than 200 µm in length) displaying a single prismatic habit or a twinned (polysynthetic or "en sablier") form ; generally, the crystals seem to grow nearly or at the contact of the stylolites (Fig. 8, C) and, in Epinoy 1 cores, their number increases with the depth : the crystals include small carbonate relics.

The growth of the albite apparently took place during the compaction from pore waters enriched by dissolution processes and the crystallization was probably favoured by thermal effect as it is suggested by Kastner (1971).

2. Silicification
Several varieties of authigenic quartz and chalcedony have been observed replacing

gypsum crystals, anhydrite and sometimes skeletal limestones. They are: single idiomorphic quartz (up to 1 cm) ; coarse quartz mosaic ; radiating fibrous quartz crystals ; fibrous minerals as quartzine and lutecite ; interlocking aggregates of idiomorphic and petaloid quartz associated with spherules or crusts of fibrous varieties. In the anhydritic nodules, the quartz encloses various quantities of anhydrite relics, sometimes so numerous that the crystal boundaries are difficult to distinguish.

In tectonic laminated anhydrite with elongated and parallel arranged crystals, the authigenic quartz encloses anhydritic relics which display an irregular or a felted texture typical of early diagenetic anhydrite (Fig. 8, D). In some Epinoy 1 samples (3135m), the silicified gypsum crystals are well preserved whereas the primary morphologies are elsewhere destroyed by the burial anhydritization and the tectonic deformation ; quartz-chalcedonic aggregates are desintegrated during the tectonic flow of the anhydrite and their fragments destroy the lamination ; sometimes (Saint Ghislain, 2,171 m) a "tail" of recrystallized calcite (pressure shadow) is developed at the border of a rotated authigenic quartz. All the observations lead to the assumption that the main part of the siliceous replacement predates the tectonic deformation. In the outcrops, replaced sulphate nodules show a brecciated texture with broken silicified fragments cemented by sparry calcite or dolomite.

Similar examples of silicified evaporites are known from formations ranging in age from Precambrian to Miocene (Munier-Chalmas, 1890 ; Siedlecka, 1972 ; Chowns and Elkins, 1974 ; Milliken 1979...) and even are considered as a memory of vanished evaporites (Folk and Pittman, 1971 ; Siedlecka 1976 ; Schreiber 1974). Silicification is a process which can take place during any time during the post-depositional history but very early growth of authigenic quartz has been reported from present day lagoonal environments (Giresse, 1968).

Fig. 8. - Accessory minerals. A. - Thin section photomicrograph of radial aggregates of lenticular crystals of celestite which are partly replaced by caltite. Saint-Ghislain borehole ; 1,787.50 m; plain light; scale bar is 200 μm. B.- Thin section microphotograph showing fluorite crystals (arrows) around a small anhydritic nodule. Epinoy 1 borehole 3,150l.62 m ; plain light; scale bar is 500 μm. C.- Albite crystals along stylolite (thin

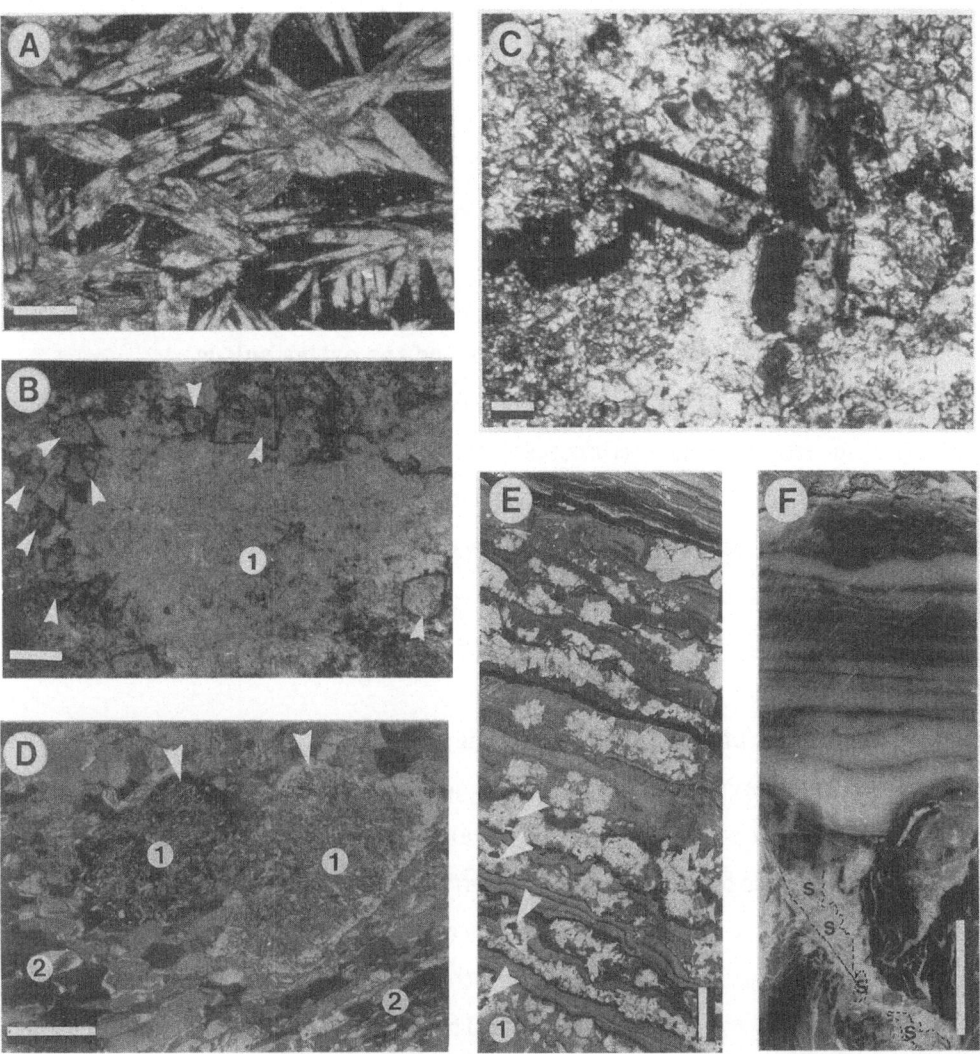

section micrograph). Epinoy 1 borehole ; 3,137.50 m ; crossed nicols ; scale bar is 20 μm.
D. - Authigenic quartz crystals (arrows) into laminated anhydrite ; in quartz (1), the
minute anhydritic inclusions develop a felted structure fairly different of the reoriented
fabric of the anhydritic host-sediment (thin section photomicrograph). Saint-Ghislain
borehole ; 2,046.60 m, scale bar is 500 μm. E. - General view of calcite pseudomorphs
after lenticular crystals of gypsum ; this transformation related to reduction of sulfates
is followed by a volume reduction (arrows show voids) and in some cases by the
formation of native sulphur illustrated in F. Saint-Ghislain borehole, 1,932.80 m ; scale
bar is 2 cm. F. - Large native sulphur pockets (S) included in the secondary calcite
infilling the fractures in a black limestone fragment. This fragment appears isolatedinto
the tectonic laminated anhydrite. Epinoy 1 borehole ; 2,937.35 m ; scale bar is 5 cm.

In our examples, an early diagenetic replacement must be considered. The silicification does not involve the external contribution of silica the origin of which can be the concentrated brines or fluids enriched by solution of siliceous organisms accumulated during the pre- or the interevaporitic sedimentation; destruction of clays minerals in relation with pH changes during diagenesis in evaporites may contribute to provide silica (B.C. Schreiber, oral comm.).

3. The carbonate replacement of sulfates and the native sulfur.

The calcite and dolomite ubiquitously replace the sulfates (gypsum, anhydrite and celestite) in various kinds of processes ; this replacement is sporadically observed at various depths in subsurface whereas the carbonate or carbonate-siliceous pseudomorphs after scattered sulfates nodules or crystals occur commonly at specific stratigraphic levels in Visean outcrops (Hennebert and Hance, 1980 ; Poels and Preat, 1983 ; Swennen *et al*, 1981 ; Swennen et Viaene, 1986 ; Rouchy, 1986).

- **Reduction of the sulfates and the native sulfur formation.** - A typical example shows centimeter-sized lenticular gypsum disseminated in black laminated limestones probably algal in origin (see above) (Saint-Ghislain, 1932 m; Fig. 8, E). The crystals are replaced by a mosaic of clear sparry calcite. Many replaced gypsum crystals have a central cavity delineated by the projecting heads of the calcite crystals. The origin of the replacement by bacterial reduction of the sulfate in organic-rich sediments is both supported by the petrographic data and by the isotopic composition of the carbon (C=-8,5 ‰, Pierre, 1986). Generally, the reaction involves : 1) a volume decrease (20 % for anhydrite-calcite, 50 % for gypsum-calcite) which can explain the holds observed in the former gypsum crystals (secondary porosity) and the diagenetic fracture of the host-sediment ; 2) the release of H_2S which can be, if it is in contact with dissolved oxygen, reoxydized into native sulfur *in situ* or in adjacent layers ; 3) production of energy. These processes are well illustrated in the Permian Castile Formation of the Texas (Kirkland and Evans, 1976 ; Shearman, 1971), in Miocene of Sicily (Dessau *et al*, 1962) and of Egypt (Rouchy *et al*, 1985).

Native sulfur nodules (cm to dm) are frequently observed in Saint-Ghislain cores and are common in Epinoy 1 where they are located within the intensely fractured (tectonic) carbonates and anhydrite (Fig. 8, F) ; the nodules are associated with the sparry calcite infilling of the fractures. In the strongly tectonized section of Epinoy 1 the native sulfur formation could be produced by reduction processes related to the hydrocarbon migration in a fractured formation ; we cannot exclude however the desulfuration of sulfurous hydrocarbons.

- Other modalities of carbonate replacement after sulfate. - Except the above described mechanism, this replacement is usually the result of the sulfate dissolution by bicarbonate waters followed by recrystallization of calcite or dolomite. Different varieties of sulfate appear to be replaced in Saint-Ghislain borehole and in Visean outcrops: lenticular gypsum, nodular or mosaïc anhydrite, single or aggregated crystals of celestite. The replacement carbonates are schematically of two distinct types :

1) aggregates of closely intergrown (30-100 μm) or mosaic of subhedral large crystals (up to 500μm) either containing more or less numerous inclusions of opaque materials and small anhydritic relics ; rarely (Yves Gomezée wells) calcite crystals display an elongate and rectangular shape and a felted fabric which suggest an isomorphic replacement of anhydrite, the relics of which can bè observed. In Saint-Ghislain well (4,161.10 m) a group of celestite crystals is replaced by an admixture of calcite, quartz and albite suggesting a relation with burial conditions. When any sulfate is preserved within the carbonate, the authigenic quartz, rich in anhydrite relics, authentifies their former sulfate nature (outcrops of Walhorn in the synclinorium of Dinant and Napoleon in the Boulonnais for instance).

2) Mosaic of large and limpid sparry calcite or dolomite (up to 1 cm); the replacement appears often in the outcrops and in the near-surface well samples (Yves

Gomezée for instance); when the two kinds of fabric are represented, the first forms an irregular outer part or floating aggregates within the second. It is assumed that the first type results from a subconcomitant dissolution-recrystallization process occurring at various stages of diagenesis of uncertain timing (early diagenesis, burial, weathering in surface conditions...), the second one is a void filling after complete dissolution of residual sulfate. Swennen *et al* (1981) described a carbonate siliceous replacement of sulfate nodules in which the quartz is the first replacement phase and the dolomite formed later after the dissolution of the residual sulfate inducing an internal brecciation. The isotopic composition of the calcite (^{13}C and ^{18}O) of the sulfate samples studied by Pierre (1986) shows two distinct groups which could be related to the surface conditions or to the fresh water phreatic zone (early or later diagenesis) and to the burial diagenesis.

3) The replacement and void-filling anhydrite.

Replacement and void-filling anhydrite are common diagenetic features in carbonates associated with evaporite layers of different ages (Dunham, 1948 ; Kendall and Walters, 1978 ; Jacka and Franco, 1974). A recent and well documented description and discussion have been presented with regard to Purbeckian beds of Aquitaine in France Clark and Shearman, 1980). The Givetian and Visean subsurface formations of Northern France and Belgium exhibit a wide variety of replacement anhydrite the study of which brings new views on the chronology of the diagenetic changes (Rouchy *et al*, 1984 a/b ; Rouchy, 1986).

There are four fundamentally different types of replacement anhydrite :

1. - The porphyroblasts (Fig. 9,A,B) are euhedral to subhedral; equant prismatic crystals (100 μm to 1 cm) generally displaying a square, triangular or rectangular section with sometimes, curved or "corroded" faces ; characteristic brown color results from the inclusion of very abundant, small relics of the host-limestone and probably of organic matter (Fuller, 1956, *in* Clark and Shearman, 1980) ; the square sections may possibly be confused with halite pseudomorphs. A narrow inclusion-free rim outlines the edges of the crystals allowing, in some cases, one to distinguish the inclusion-rich

crystalline body from its matrix (in thin section.) Crystals occur singly, in association of a few individuals or in polycrystalline aggregates in which the crystals loose their euhedral morphology. It is very important to note that porphyroblasts (and the veinlets) grow preferentially in the fine-grained limestones.

2. - The porphyroblasts are frequently associated with anhydritic veinlets according to two principal fashions (Fig. 9C): first, the porphyroblasts are arranged on both sides of the veinlet which is filled with clear anhydrite, suggesting the replacement progresses into the carbonates from the veinlets which can be little fractures or other discontinuities (bioturbation); second, some veinlets are composed of clear anhydrite in a narrow axial zone,outlined by an inclusion-rich rim. When clear anhydrite-filled veinlets cut across the porphyroblasts, the common optical characters suggest that the porphyroblasts and the veinlet form a single crystal and have the same origin ; this observation is in agreement with Clark and Shearman's interpretation assuming the veinlets were the expression of the volume increase (25 %) generated by the replacement itself.

3. - The large monocrystals, termed domino-like or stairstep (Dunham, 1948; Jacka, 1977), differ from the preceding by an irregular outline with rectangular-shaped projections and re-entrants. (Fig. 9D). Contrary to the porphyroblasts, this kind of replacement of anhydrite is preferentially developed in any matrix such as oolitic, peloidal or skeletal limestones. This difference in behaviour is not well understood. The structure of the host-limestone components (oolits, pellets, shells) may be cut linearly by the boundary of the crystal and is faithfully revealed within the anhydrite by numerous relics. Large anhedral anhydrite crystals form a pseudo-cement filling the interoolite porosity and partially replacing the components; these crystals differ from the preceding by the lack of rectilinear boundaries, probably due to a competitive growth of the adjacent crystals. Several thin sections of peloidal limestone of Epinoy 1 show features as the simultaneous replacement of several peloids and of some part of the cement by the same crystal of anhydrite (Fig. 9,E).

4. - Some rounded or oblong bodies composed of radiating or interlocked aggregate of large bladed crystals (up to 2 cm) seem to have filled cavities of organic or of

dissolution origin in the limestone ; similar aggregates form the cement of breccia (Fig. 9, F). Nevertheless, the conservation of carbonate relics floating into the anhydrite crystals seems to indicate that if the void-filling is the principal control, a concomitant replacement must be considered.

The relationships between the replacement anhydrite and the structure of the host-sediment reveal the substitution post-dates compaction and the interlocking of the ooids. More significant are the relations with tectonic deformation. At times the porphyroblast aggregates are developed in bends of microfolds in the limestones but they are not themselves deformed by the folding. The growth of the replacive porphyroblasts appears to be initiated along the block joins which constitute glide planes. Besides, the associated veinlets cut the axial phase of the microfold obliquely. Other examples from Saint-Ghislain core show large porphyroblasts growing from stylolites ; similar features are also observed in Triassic of the Jura (France, Pisu and Rouchy, unpublished report).

All the observations, taken together, allow the dating of the replacement as later than limestone diagenesis and in many examples, synchronously or later than the tectonic deformation ; another example of late diagenetic control is illustrated by Kendall and Walters (1978) in the Mississipian limestones in which the anhydrite substitution appears post-Triassic in age. Nevertheless, it is evident this dating cannot be applied to the anhydrite replacement in all formations. Clark and Shearman (1980) have observed, for instance in Purbeckian and in Devonian beds, replacements formed before the first phase of limestone diagenesis.

Fig. 9. - Microscopic view of diagenetic replacements. A. - Anhydrite porphyroblasts in the hinge of a microfold; each porphyroblasts contains numerous micritic impurities of the host sediment and a narrow clear border. Note also the anhydritic veinlets which cut obliquely the microfold axis; Saint-Ghislain borehole; 2,036.90 m; plain light; scale bar is 500 µm. B. - Porphyroblasts surrounding nodular reoriented anhydrite indicating their growth post-dates the deformation. Saint-Ghislain borehole; 2,041.78 m; plain light ; scale bar is 500 µm. C. - Large replacement anhydrite veinlets similar to the porphyroblasts in having inclusions-rich body, narrow clear border and limpide anhydrite in the middle part ; these veinlets are frequent in the more deformed part of the Epinoy 1 formation. Epinoy 1 borehole; 3,148.02 m; plain light; scale bar is 1 mm. D. - Domino-like replacement anhydrite (clear) in a peloidal limestone (dark). Saint-Ghislain.

55

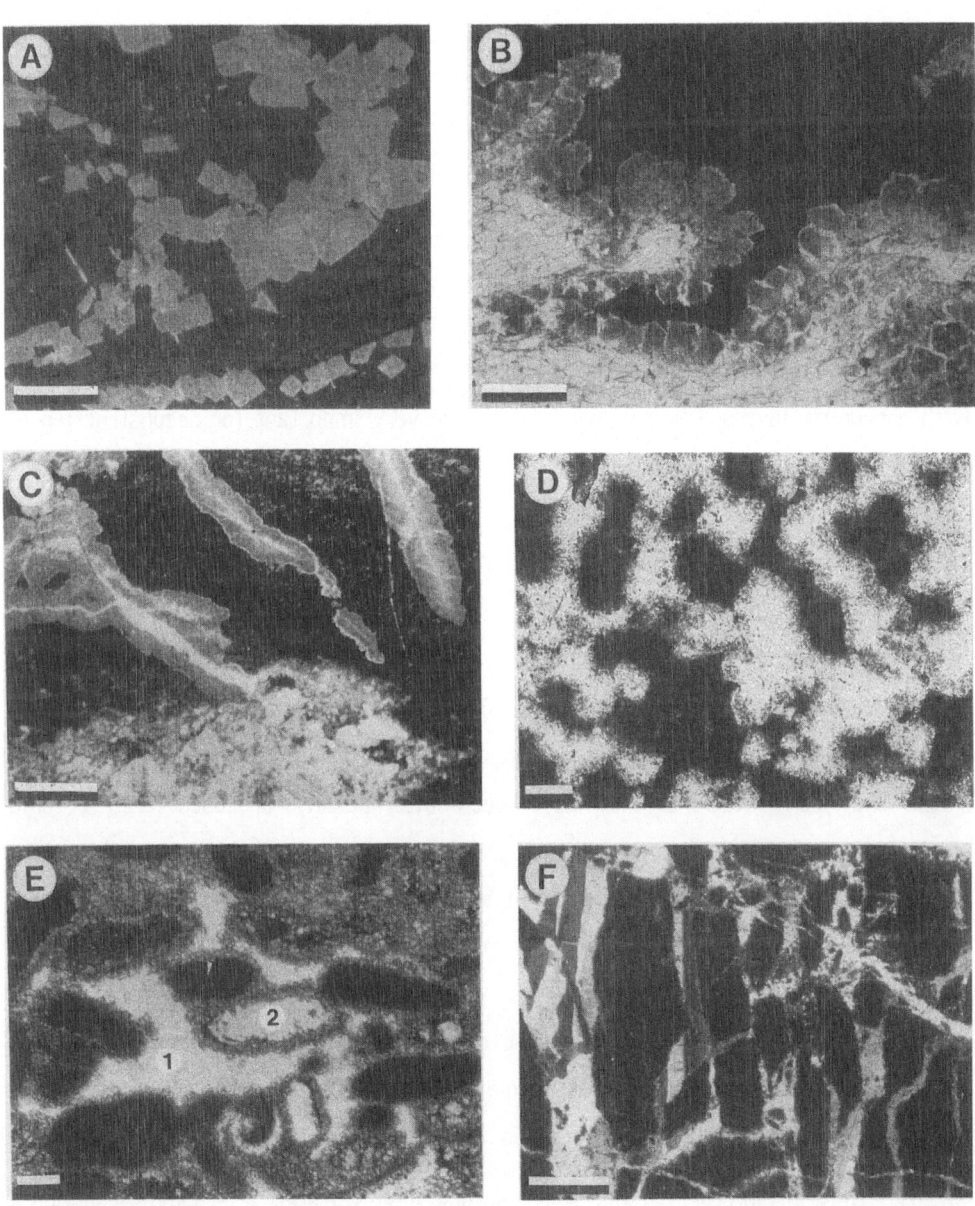

borehole; 2,054.30 m; plain light; scale bar is 100 μm. E. – Monocrystalline anhydrite (clear) replacing the cement (1) as well as the peloids (2) in a peloidal limestone. Epinoy 1 borehole; 3,135.30 m; plain light; scale bar is 200 μm. F. – Prismatic anhydrite cementing a carbonate breccia. Saint Ghislain borehole; 2,057.53 m; crossed nicols ; scale bar is 500 μm.

Probably, the anhydritic substitution implies a nearly concomitant dissolution of the limestone and precipitation of anhydrite, processes which can be controlled by the migration of fluids along the diagenetic or tectonic discontinuities (stylolites, small gliding planes, bends of microfolds, anhydrite-limestone boundary...) and by structural opening of porosity.

Even if the replacive anhydrite in the limestone does not imply that the limestones were deposited in evaporitic environments, as pointed out by Clark and Shearman (1980), their presence provides the evidence that these limestones are or have been interbedded with anhydritic layers. Their recognition can be very important for reconstruction of vanished evaporitic formation (Rouchy *et al*, 1986b).

THE BRECCIAS

One of the more difficult geological problems in Belgium concerns the genesis of the Visean great breccias, particularly the "Grande Brèche de Namur et de Dinant", which end the Visean sequence throughout the greater part of the Dinant and Namur structural units. The interpretations that have been alternatively proposed are tectonic fragmentation, sedimentary transport or gravity flow (olistostroms), favoured by the presence of evaporites as postulated by Pirlet and Bouckaert, 1976). The various facies of the brecciated horizons have been documented by Bourguignon (1950-1951) in his synthetic work. West (1969) was the first who indicate, in an unpublished report, the presence of gypsum pseudomorphs in the cement of the "Grande Brèche". Since the discovery of Saint-Ghislain evaporitic formation, many of the studies focused on Visean Limestone outcrops showed the numerous occurrences of pseudomorphosed sulfates,

Fig. 10. - Dissolution breccias and pseudomorphs after sulfates. A. - Sample showing a calcite pseudomorph after mosaic anhydrite; Salet road near Dinant, Dinant nappe. Scale bar is 2 cm. B. - Thin section microphotograph showing sparry calcite pseudomorphs after sulfates probably crystalline aggregates of gypsum. Yves Gomezée borehole, S2 ; 26.50 m; plain light; scale bar is 1 mm. C. - Thin section photomicrograph showing domino-like anhydrite (white) replaced by sparry calcite. Note the characteristic outline

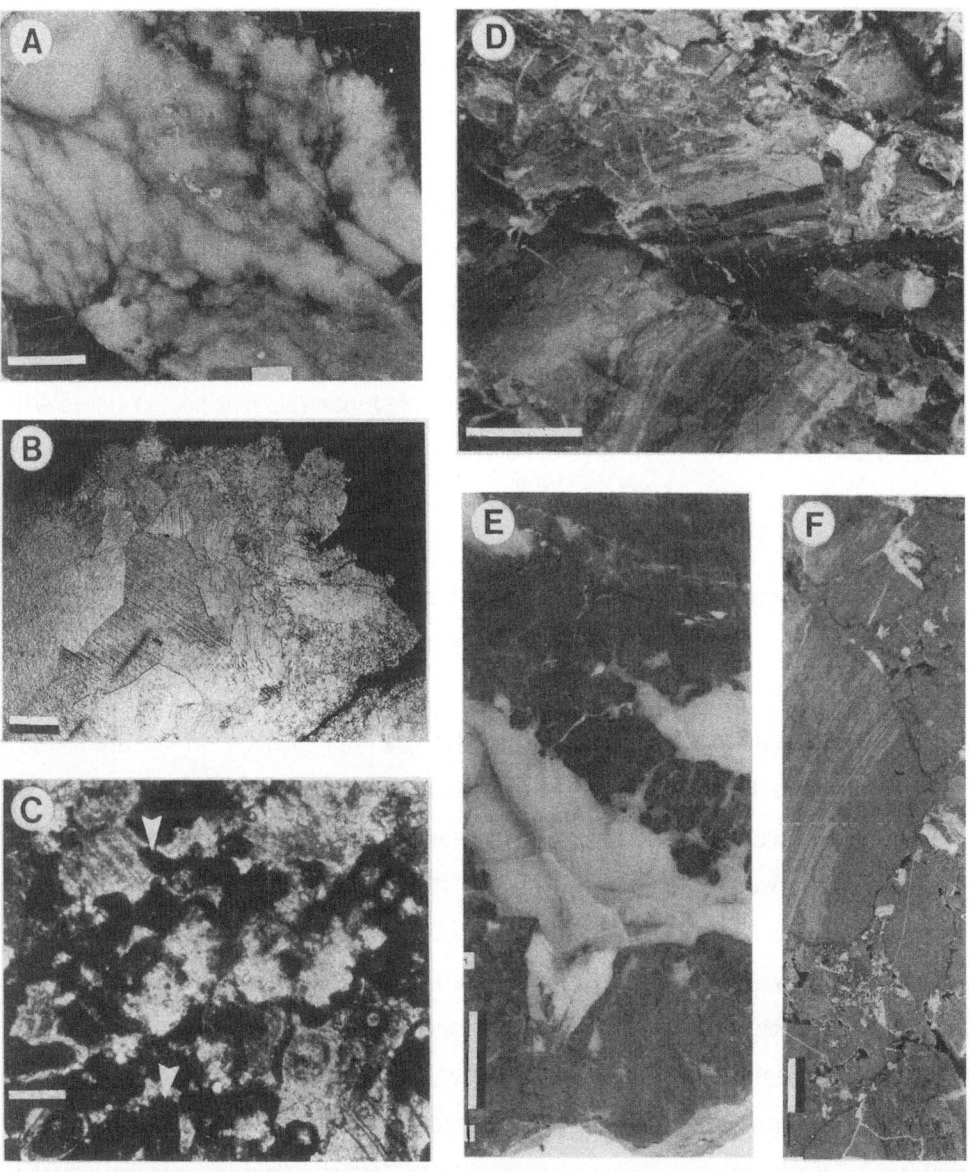

anhydrite with square pattern, re-entrants and projections ; we can recognize pellets (arrow) replaced into former anhydrite. Yves Gomezée boreholes, S8; 12.70 m ; scale bar is 500µm. D. - Breccia probably related to the sulfates dissolution. Quarry of Landelies, Dinant nappe. Scale bar is 10 cm. E. - Mechanical breccia with anhydritic cement. Epinoy 1 borehole; 3,144.10 m; scale bar is 2 cm. F. - Solution-breccia in the deep karst of Saint-Ghislain. Saint-Ghislain borehole; 2,529.45 m; scale bar is 2 cm.

particularly in association with brecciated horizons (Fig. 10, A, B, C). These observations lead to the conclusion the breccia have been originated from collapse after the evaporite solution (Swennen *et al*, 1981; Swennen and Viaene, 1986; Mamet *et al*, 1986; Rouchy *et al*, 1986a/b).

The breccia beds can be encountered in subsurface (Douvrain, Ghlin, Wépion, Saint Ghislain) but the extensive formations appear in the surface outcrops or at shallow depth as, for instance, in the Yves Gomezée wells near Philippeville (Rouchy *et al*, 1986b). Many authors (Bless *et al*, 1980; Groessens *et al*, 1979; Rouchy *et al*, 1986a; Conil and Groessens 1985) have pointed out that the brecciated intervals may be correlated with bands of anhydrite in the Saint-Ghislain borehole; the V2b-V3a interval (Middle Visean) contains both the upper massive anhydrite of Saint Ghislain and the great breccia of Namur and Dinant.

Thus, in the Douvrain borehole which is situated about 4 kilometers of Saint-Ghislain, the upper part of the evaporite section is represented by a breccia associated with silicified sulfates (Leclercq, 1980).

The cores of the Yves Gomezée wells and the quarry of Landelies in the Dinant nappe offered an excellent opportunity to study the great breccia of the V2b-V3a (Rouchy *et al*, 1986b) and to elucidate the timing of the dissolution. The fragments are highly angular and irregular as well in shape as in size and the fabric is often chaotic (Fig. 10, D) ; almost continuous layers of limestones are floating within the breccia. The matrix type varies irregularly : white recrystallized calcite, silt, very fine calcite, etc. Similar features have been observed in karstified evaporites of Messinian age or in Triassic "Calcare cavernoso" from Tuscany (B.C. Schreiber, oral comm.).

The fragments as well as the unbrecciated carbonates, contain nodular sulfates replaced by white calcite ; some of them may be former anhydritic nodules whereas others which have rectilinear and angular boundaries are probably pseudomorphosed gypsum aggregates (Fig. 10, B). The microfabric of carbonates (see above) and the preservation of minute anhydrite relics in the calcite and in the scattered authigenic quartz make obvious the replacement; pseudomorphs after lenticular gypsum are often observed.

The recognition of the pseudomorphs after replacement anhydrite in the form of domino-like monocrystals (Fig. 10,B), porphyroblasts and veinlets, corroborates the former presence of evaporite interbeds indicating that the brecciation post-dates the burial diagenesis and even probably one stage of tectonic deformation. In the cores of Yves Gomezée wells and in Landelies outcrops, the fragments present a network of fractures truncated at the fragment boundaries (Fig. 10,D). This observation confirms that in this example, the brecciation clearly seems to have taken place after a tectonic fragmentation (Fig. 10,E). It is probable that the extensive solution of the evaporites begins with the Permian denudation and continues over the long period until recently, as is postulated by Delmer *et al* (1982) and de Magnée *et al* (1986). It is assumed here that a mechanical pre-brecciation and the solution processes favour the increase of water circulation. Thus, the presence of a saline water aquifer in the deep cave solution with breccia (Fig. 10,F) at the base of the anhydritic formation on the Saint Ghislain well (Delmer *et al*, 1982) is very demonstrative ; this cave coincides probably with a minor tectonic glide plane (Rouchy *et al*, 1984 a/b).

It is not suggested however that all the Visean breccias are necessarily formed in this way, other breccias could have a different timing. In the breccia of Namur (Grands Malades), Mamet *et al* (1986) provide criteria of early polyphased brecciation.

On the basis of sedimentological and stratigraphical data, the formation of the Great Visean breccia is most logically explained by collapse after dissolution of anhydrite (eventually salt) interbeds of significant thickness; thus, the extensive distribution of the breccia and of the pseudomorphs which cover a large part of the synclinorium of Namur and of the nappe of Dinant, provides the evidence of southward extension of the evaporites; their original distribution appears to have been independant of the present organization in the structural units.

CONCLUSION

The complicated distribution of the Visean evaporites in north-western Europe

(Northern France and Belgium) is inherited from a complicated paleogeographic, tectonic and post-tectonic history which have strongly modified their former facies, thickness and limits. The stratigraphical and sedimentological studies of the thick anhydritic deposits, as well as the pseudomorphs and breccias in the outcrops, allow to the reconstruction of the depositional modalities, the successive diagenetic changes, the deformational features and the post-tectonic events.

Diversified environments of deposition resulting from repeatedly restricted open lagoonal conditions led to the deposition of subaqueous sulfates (gypsum) and sporadic subaerial anhydrite diagenesis; in the thickest formations, the predominant nodular and mosaic structures is interpreted as resulting of burial conversion of gypsum to anhydrite which stresses the primary depositional features rather than a generalized early diagenesis in sabkha-like conditions.

The chronology of the mineralogical and textural post-sedimentary changes was established based on the early diagenesis (celestite and fluorite authigenesis, silicifications, limited gypsum/anhydrite conversion), the burial conditions (complete gypsum-anhydrite conversion, albite authigenesis, sulfate calcitization...), and the tectonic deformation (carbonates replaced by anhydrite during or immediately after the deformation).

The deformational fabrics of the anhydrite in relation with Hercynian tangential stresses and subsequenf flow mechanisms, leads to the destruction of the primary structures and to the development of structures sometimes similar to some metamorphic rocks (stretching, boudinage, microfolding, tectonic lamination or schistosity, augen-like, mylonitic). Some of the tectonic features may mimic sedimentary structures. The recognition of the deformational features allows us to envisage the important role of the evaporites in the Hercynian deformations. The evaporites supplied detachment and gliding planes and favoured formation of minor slice thrusts; this is suggested for the base of anhydrite formation of Saint-Ghislain and demonstrated by the implication of Epinoy 1 evaporites in reverse position and in a multi-system of slices below the major overthrust of the Midi.

The formation of the greatest part of the extensive Visean breccia ("Grande Brèche de Namur et de Dinant") may be most logically explained by solution-collapse as suggested by the presence of pseudomorphs of evaporites, fabrics of the breccia, stratigraphic correlation between breccias and thick anhydritic formation...

It is assumed the dissolution occurred after the Hercynian deformation and in some cases, until a recent period (cf. breccia of Saint-Ghislain) ; this observations lead to the conclusion that widespread evaporitic body of Visean age (Middle Visean, V2b-V3a at least) extended into the Dinant and Namur structural units.

Although the area in which evaporation and precipitation took place cannot be exactly delineated in geographic extent, all the data provide the evidence that the isolated thick anhydritic deposits (Saint-Ghislain and Epinoy 1) represent the relics of more widespread evaporitic formation extending more or less throughout the different structural units: autochthonous in the Namur basin, para-autochtonous, allochtonous in the Dinant nappe. Its present discontinuity is due to structurally and halokinetically controlled thickening in some areas and, in contrast, the thinning or the disappearance in others; this latter results from the combination of mechanical lamination and breaking and/or of dissolution.

The important role of the evaporites in the genesis of the Hercynian regional setting which displays an Appalachian thin-skinned type of deformation appears well documented both by the deformational fabrics of the anhydrite and by the structural framework of Saint-Ghislain and Epinoy 1 formations.

ACKNOWLEGMENTS. - We are grateful to C. Pierre for her active participation to this work studying the isotopic composition of the Visean sulfates and carbonates described in this paper and for the stimulating discussions, B. Moine and D. Couilloud so have provided many petrographical and geochemical indications. The authors thank B.C. Schreiber for reading the manuscript and offering valuable suggestions for its improvement.

Subsurface cores and logs have been provided by the Geological Survey of Belgium and the Compagnie Française des Pétroles TOTAL.

Research was conducted on a grants from the Compagnie Francaise des Pétroles TOTAL and from the National Programme "Géologie Profonde de la France".

Gratitude is also extended to Mrs A. Cambreleng, N. Day, R. Deletoille, M. Destarac, G. Tortel, Mr L. Ganon for the quality of their assistance in manuscript drafting, reading, typing, photography and thin section making.

BIBLIOGRAPHY

Aubouin J., 1985. - Géophysique. Le programme E.C.O.R.S. Encyclopaedia Universalis, Universalia 1985, p. 273-277.

Becq-Giraudon J.F., Colbeaux J.P. & Leplat J., 1981. - Structures anciennes transverses dans le bassin houiller du Nord-Pas-de-Calais. Ann. Soc. géol. Nord., C I, p. 117-123.

Becq-Giraudon J.F., 1983. - Synthèse structurale et paléogéographique du bassin houiller du Nord. Mém. B.R.G.M., n° 123, 68 p.

Bless J.M., Conil R., Defourny P., Groessens E., Hance L. & Hennebert M., 1980. - Stratigraphy and thickness variations of some Struno-Dinantian deposits around the Brabant massif. Meded. Rijks Geol. Dienst., 32-1, p. 56-65.

Bless J.M., Boonen P., Bouckaert J., Brauckmann C., Conil R., Dusar M., Felder P.J., Felder W.M., Gokdag H., Hockel F., Laloux M., Langguth R.H., Van Der Mer Mohr C.G., Meessen J.P.M.TH., Ophet Veld F., Paproth E., Pietzner H., Plum J., Poty E., Scherr A., Schulz R., Streel M., Thorez J., Vanroijen P., Vanguestaine M., Vieslet J.L., Wiersma D.J., Winkler Prim C.F. & Wolf M., 1981. - Preliminary report on Lower Tertiary-Upper Cretaceous and Dinantian-Famennian rocks in the boreholes Heugem A and Kastanjelaan 2 (Maastricht, the Netherlands). Meded. Rijks Geol. Dienst., 33-13, p. 333-415.

Bouquillon A., 1984. - Stratigraphie, paléoenvironnement et diagénèse dans le Primaire sédimentaire des forages du Nord de la France. Mém. D.E.A., Université de Lille, 53 p

Bourguignon P., 1950-1951. - Etude géologique et sédimentologique des brèches calcaires viséennes de Belgique. Ann. Soc. géol. Belgique, 74, 3, p. 105-211.

Cazes M, Torreilles G., Bois C., Damotte B., Galdeano A., Hirn A., Mascle A., Matte Ph., Pham Van Ngoc & Raoult J.F., 1985. - Structure de la croûte hercynienne du Nord de la France : premiers résultats du profil ECORS. Bull. Soc. géol. Fr., 8ème sér., I, 6, p. 925-941.

CFP(M), COPESEP, RAP & SNPA., 1965. - Contribution à la connaissance des bassins paléozoïques du Nord de la France, Ann. Soc. Géol. Nord., LXXXV, p. 273-281.

Chowns T.M. & Elkins J.E., 1974. - The origin of quartz geodes and cauliflower cherts through the silicification of anhydrite nodules. J. Sedim. Petrol., 44, 3, p. 885-903.

Ciarapica G. & Passeri L., 1976. - Deformazioni da fluidificazione ed evoluzione diagenetica della formazione evaporitica di Burano. Boll. Soc. geol. Ital., 95, 5, p. 1175-1199.

Clark D.N., & Shearman D.J., 1980. - Replacement anhydrite in limestones and the recognition of moulds and pseudomorphs; a review. Rev.

Inst. Invest. Geol. (Diputacion Provin. de Barcelona), 34, p. 161-186.

Coen-Aubert M., Groessens E. & Legrand R., 1980. - Les formations paléozoïques des sondages de Tournai et Leuze. Bull. Soc. belge Géol., 89, 4, p. 241-275.

Colbeaux J.P., Beugnies A., Dupuis C., Robaszinsky F. & Somme J., 1977. - Tectonique de blocs dans le Sud de la Belgique et le Nord de la France. Ann. Soc. géol. Nord., XCVIII, p. 191-222.

Conil R. & Groessens E., 1985. - La place des évaporites dans l'échelle stratigraphique du Dinantien. Coll. "Evaporites pré-permiennes en Europe" (Bruxelles, 9-10 mai 1985), GRECO 52, Société belge de Géol. et Groupe de Contact Sédimentol., abstract, 2 p.

Dejonghe L., Delmer A. & Groessens E., 1976. - Découverte d'anhydrite dans les formations anténamuriennes du sondage de Saint-Ghislain. Bull. Acad. Roy. Belg. (Cl. Sci.), Séance 10.1.1976, p. 80-83.

Delmer A., 1972. - Origine du bassin crétacique de la vallée de la Haine. Serv. géol. Belg., Prof. Paper, 1972/5, 13 p.

Delmer A., 1977. - Le bassin du Hainaut et le sondage de Saint-Ghislain. Serv. géol. Belg., Prof. Paper, 1977/6, n°143, 12 p.

Delmer A., Leclercq V., Marlière R. & Robaszinsky F., 1982. - La géothermie en Hainaut et le sondage de Ghlin (Mons, Belgique). Ann. Soc. géol. Nord CI (1981), p. 189-206.

De Magnée I., Delmer A. & Cordonnier M., 1986. - La dissolution des évaporites du Dinantien et ses conséquences. Bull. Soc. belge Géol., 95, 2-3, p. 213-220.

Dessau G., Jensen M.L. & Nakai N., 1962. - Geology and isotopic studies of silician sulfur deposits. Econ. Geol., 57, p. 410-438.

Dunham K.C., 1948. - A contribution to the petrology of the Permian evaporites deposits of northeastern England. Proc. Yorkshire Geol. Soc., 27, p. 217-227.

Folk R.L. & Pittman J.S., 1971. - Length slow chalcedony ; a new testament for a vanished evaporites. J. Sedim. Petrol., 41, 4, p. 1045-1058.

Georges T.N., 1963. - Tectonic and palaeogeography of the British Isles. Proc. Yorkshire Geol. Soc., 31, p. 227-318.

Giffard H.P.W., 1922-1923. - The recent search for oil in Great Britain. Trans. Inst. Min. Engrs., 65, p. 221-250.

Giresse P., 1968. - Authigenèse actuelle de quartz pyramidés dans la lagune de Fernan Vaz (Gabon). C. R. Acad. Sci. Paris, 267, sér. D, p. 145-147.

Goemaere E., Thorez J., & Dreesen R., 1985. - A propos des milieux évaporitiques supratidaux dans les psammites du Condroz (Famennien supérieur, Belgique). Coll. "Evaporites pré-permiennes en Europe" (Bruxelles, 9-10 mai 1985) Greco 52, Soc. belge Géol. et Groupe de Contact Sédimentol., abstract., 2 p.

Graulich J.M., 1963. - Les résultats du sondage de Soumagne. Serv. géol. Belg., Prof. Paper, 1977/2, 55p.

Groessens E., Conil R. & Hennebert M., 1979. - Le Dinantien du sondage de

Saint-Ghislain. Stratigraphie et Paléontologie. Mém. Expl. Cartes géol. min. Belgique, 22, 137 p.

Guelorget D. & Perthuisot J.P., 1983. - Le domaine paralique; expressions géologiques, biologiques et économiques du confinement. Trav. Lab. Géol. Ecole Normale sup. Paris., 136 p.

Hance L. & Hennebert M., 1980. - On some lower and middle Visean carbonate deposits of the Namur basin, Belgium. Meded. Rijks Geol. Dienst, 32.9, p. 66-68.

Hennebert M. & Hance L., 1980. - Présence de nodules de sulfate de calcium silicifiés dans le Viséen moyen (cf. V2b) à Vedrin (Namur, Belgique). Ann. Soc. géol. Belgique, 103, p. 25-33.

Helman M.L. & Schreiber B.C., 1983. - Permian Evaporite Deposits of the Italian Alps (Dolomites) : The development of Unusual and Significant Fabrics. Sixth Intern. Symp. on Salt, Salt Institute, p. 57-66.

Jacka A.D., 1977.- Deposition and diagenesis of the Fort Terret formation (Edwards Group) in the vicinity of junction, Texas. In: Cretaceous Carbonates of Texas and Mexico. Bebout D.G. and Loucks R.G., eds. Texas Bureau of Economic Geology, 89, 182-200.

Jacka A.D. & Franco L.A., 1974. - Deposition and Diagenesis of Permian Evaporites and Associated Carbonates and Clastics on Shelf Areas of the Permian Basin. Fourth Intern. Symp. on Salt., North. Ohio Geol. Soc., 1, p. 67-89.

Kastner M., 1971. - Authigenic feldspars in carbonate rock. Amer. Miner., 56, p. 1403-1442.

Kendall A.C. & Walters K.L. 1978. - The age of metasomatic anhydrite in Mississipian reservoir carbonates, southeastern Saskatchewan. Canad. J. Sc., 15, 3, p. 424-430.

Kinsman D.J.J., 1969. - Interpretation of Sr^{++} concentrations in carbonate minerals and rocks. J. Sedim. Petrol., 39, 2, p. 486-508.

Kirkland D.W. & Evans R., 1976. - Origin of limestones buttes, Gypsum Plain, Culberson Country, Texas. Am. Assoc. Petrol. Geol. Bull., 21, p. 833-898.

Laumondais A., Rouchy J.M. & Groessens E., 1984. - Importance des formations anhydritiques dinantiennes pour l'interprétation paléogéographique et structurale du domaine varisque d'Europe septentrionale. C. R. Acad. Sci. Paris, 298, II, 9, p. 411-414.

Leclercq V., 1980. - Le sondage de Douvrain. Serv. géol. Belgique, Prof. Paper, 1980/3, 51 p.

Llewelyn P.G. & Stabbins R., 1968. - Core anhydrite from the Anhydrite Series, carboniferous, p. 171-186.

Llewelyn P.G., Mahmoud S.A. & Stabbins R., 1968. - Nodular anhydrite in Carboniferous Limestone, Hathern borehole, Leicestershire, West Cumberland. Trans. Inst. Min. Metall., B, 77, p. 21-25.

Loucks R.G. & Longman M.W., 1982. - Lower Cretaceous Ferry Lake

Anhydrite, Fairway Field, East Texas ; product of shallow-subtidal deposition. *In* C.R. HANDFORD *et al.* (ed.): Depositional and Diagenetic Spectra of Evaporites. A core workshop, S.E.P.M. core workshop n° 3, Calgary, p. 130-173.

Lowenstam H.A., 1981. - Minerals formed by organisms. Science, 211, p. 1126-1131.

Mamet B., Claeys P., Herbosch A., Preat A. & Wolfowicz P., 1986. - La "Grande Brèche" viséenne (V3a) des bassins de Namur et de Dinant (Belgique) est probablement une brèche d'effondrement. Bull. Soc. belge de Géol., 95, 2-3, p. 151-166.

Martinez J.D., 1974. - Tectonic behavior of evaporite.s *In* A.H. COOGAN (ed.) Fourth Intern. Symp. on Salt, North. Ohio Geol. Soc., 1, p. 155-168.

Milliken K.L., 1979. - The silicified evaporite syndrome. Two aspects of silicification history of former evaporite nodules from the Southern Kentucky and northern Tennessee. J. Sedim. Petrol., 41, 1, p. 245-256.

Müller W.H., Schmid S.M. & Briegel U., 1981. - Deformation experiments on anhydrite rocks of different grain sizes ; rheology and microfabric. Tectonophysics, 78, p. 527-543.

Munier-Chalmas E., 1890. - II. Sur les dépôts siliceux qui ont remplacé le gypse. C. R. Acad. Sci. Paris, 110, p. 663-666.

Murray R.C., 1964. - Origin and diagenesis of gypsum and anhydrite. J. Sedim. Petrol., 34, p. 512-523.

Pierre C., 1986. - Données de géochimie isotopique sur les anhydrites (^{18}O, ^{34}S) et les carbonates diagénétiques (^{18}O, ^{13}C) des séries évaporitiques givétiennes et viséennes du Nord de la France et de la Belgique. Bull. Soc. belge Géol., 95, 2-3, p. 129-138.

Pierre C., Rouchy J.M., Laumondais A. & Groessens E., 1984. - Sédimentologie et géochimie isotopique (^{18}O, ^{34}S) des sulfates évaporitiques givétiens et dinantiens du Nord de la France et de la Belgique ; importance pour la stratigraphie et la reconstitution des paléomilieux de dépôt. C. R. Acad. Sci. Paris, 299, II, 1, p. 21-26.

Pierre C., & Rouchy J.M. 1986. - Oxygen and sulfur isotopes in anhydrites from Givetian and Visean evaporites of Northern France and Belgium. Chem. Geol. (Isotope Geoscience section), 58, p. 245-252.

Pirlet H. & Bouckaert J., 1976. - A propos de l'âge post-namurien de la Grande Brèche de la station de Dinant. Ann. Soc. géol. Belgique, 99, p. 147-154.

Poels J.P. & Preat A., 1983. - Mise en évidence d'une série évaporitique dans le Viséen inférieur de Vedrin (Province de Namur). Bull. Soc. belge Géol., 92, 4, p. 337-350.

Preat A. & Rouchy J.M., 1986. - Faciès préévaporitiques dans le Givétien des bassins de Dinant et de Namur. Bull. Soc. belge Géol., 95, 2-3, p. 177-190.

Rouchy J.M., 1976. - Sur la genèse des deux principaux types de gypse (finement lité et en chevrons) du Miocène terminal de Sicile et d'Espagne méridionale. Rev. Géogr. phys. Géol. dyn., (2), XVIII, p. 347-364.

Rouchy J.M., 1986. - Sédimentologie des formations anhydritiques givétiennes et dinantiennes du segment varisque franco-belge. Bull. Soc. belge Géol., 95, 2-3, p. 111-128.

Rouchy J.M., Maurin A.F. & Bernet- Rollande M.C., 1980. - Méthodes de description (terrain, subsurface, laboratoire) destinée à une meilleure compréhension de la sédimentation des évaporites. In Méthode d'étude des évaporites, Ed. Technip., p. 11-28.

Rouchy J.M., Groessens E. & Laumondais A., 1984a. - Sédimentologie de la formation anhydritique viséenne du sondage de Saint-Ghislain (Hainaut, Belgique). Implications paléogéographiques et structurales. Bull. Soc. belge Géol., 93, 1-2, p. 105-145.

Rouchy J.M., Pierre C., Moine B., Couilloud D., Laumondais A. & Groessens E., 1984b. - Sédimentation, diagenèse et déformations tectoniques des évaporites paléozoïques; intérêt pour l'interprétation paléogéographique et structurale. Programme Géologie profonde de la France, 1ère phase d'investigations 1983-1984; rapports généraux et communications. Thème I: Chevauchements nord-varisques. Doc. B.R.G.M., 81-1, p. 71-82.

Rouchy J.M., Monty C., Pierre C., Bernet-Rollande M.C., Maurin A. & Perthuisot J.P., 1985. - Genèse de corps carbonatés diagénétiques par réduction de sulfates dans le Miocène évaporitique du Golfe de Suez et de la Mer Rouge. C. R. Acad. Sci. Paris, 301, 16, p. 1193-1198.

Rouchy J.M., Pierre C., Groessens E., Monty C., Laumondais A. & Moine B., 1986a. - Les évaporites pré-permiennes du segment varisque franco-belge, aspects paléogéographiques et structuraux. Bull. Soc. belge Géol., 95, 2-3, p. 139-150.

Rouchy J.M., Groessens E. & Conil R., 1986b. - Signification des pseudomorphoses d'évaporites associées aux brèches viséennes dans les sondages de Yves Gomezée (Synclinorium de Dinant, Belgique). Bull. Soc. belge Géol., 95, 2-3, 167-176.

Rouchy J.M., Bernet-Rollande M.C. & Maurin A.F., 1986c. - Pétrographie descriptive des évaporites. Applications sur le terrain, en subsurface et au laboratoire. In, Les séries à évaporites en exploration pétrolière, Tome 1 : Méthodes géologiques. Ed. Technip., p. 73-122.

Sabouraud-Rosset C., 1970. - Sur les compagnons de cristallisation du gypse. C. R. Acad. Sci., Paris, 270, D, p. 1-2.

Schreiber B.C., 1974. - Vanished evaporites : revisited. Sedimentology, 21, p. 329-331.

Schwerdtner W.M., 1966. - Intragranular gliding in domal salt. Tectonophysics, 5, 5, p. 353-380.

Schwerdtner W.M., 1974. - Schistosity in Deformed Anhydrite - A reinterpretation. Fourth Intern. Symp. on Salt, North. Ohio Geol. Soc., I, p.

235-240.

Shearman D.J., 1971. - Marine evaporites. The calcium sulphate facies. Unpubl. notebook. The Univ. of Calgary, A.A.P.G. Seminar, 65 p.

Shearman D.J., 1985. - Syndepositional and Late Diagenetic Alteration of Primary Gypsum and Anhydrite. Sixth Internat. Symp. on Salt, Salt Inst., 1, p. 41-50.

Siedlecka A., 1972. - Length-slow chalcedony and relicts of sulfates - Evidences of evaporitic environments in the Upper Carboniferous and Permian beds of Bear Island, Svalbard. J. Sedim. Petrol., 42, 4, p. 812-816.

Siedlecka A., 1976. - Silicified Precambrian evaporite nodules form Northern Norway : a preliminary report. Sedim. Geol., 16, p. 161-175.

Sonnenfeld P., 1984. - Brines and Evaporites. Academic Press, 613 p.

Swennen R., Viaene W., Jacobs L. & Van Orsmael J., 1981. - Occurrence of calcite pseudomorphs after gypsum in the Lower carboniferous of the Vesder region (Belgium). Bull. Soc. belge Géol., 90, 3, p. 231-247.

Swennen R. & Viaene W., 1986. - Occurrence of pseudomorphosed anhydrite nodules in the Lower Visean (Lower Molinacian of the Verviers Synclinorium, E. Belgium). Bull. Soc. belge Géol., 95, 2-3, p. 88-99.

Wall R.W., Murray G.E. & Diaz T.G., 1961. - Geological occurrence of intrusive gypsum and its effect on structural forms in Cohahuila marginal folded province of Northern Mexico. Am. Assoc. Petrol. Geol. Bull., 45, 9, p. 1504-1522.

West I.M., 1969. - Examination of the Grande Brèche of Belgium, a project supported by a grant from the Bernard Hobson Fund. Rep. of the British Assoc. for the Advanc. of Sci., Unpubl. Rep., 10 p.

West I.M., Brandon A. & Smith M., 1968. - A tidal flat evaporitic facies in the Visean of Ireland. J. Sedim. Petrol., 38, 4, p. 1079-1093.

DEPOSITIONAL MODELS OF LOWER AND MIDDLE TRIASSIC EVAPORITES IN THE UPPER YANGTZE AREA, CHINA

Wu Yinglin and Yan Yangji
Chengdu Institute of Geology and
Mineral Resources
Chengdu, Sichuan
China

REGIONAL AND PALEOENVIRONMENTAL SETTING

The Upper Yangtze area is located in the south-west China (Figure 1) between the latitudes of 26° and 33° north and longitudes 102° and 110° east. Its basement was formed in Precambrian during the Jinning orogenic movement (1.5 b.y.). The sedimentary cover (Sinian to Permian) is about 2,000 m thick. During Triassic the area considered was near the equator and located east of the Paleotethys. The correlation of the Triassic between the west, middle and east regions are shown in Table 1.

During Lower Triassic, in the west margin of the platform there were coastal mountains with the north-south strike. At the beginnig of Lower Triassic the Upper Yangtze area basically was an open sea, then the oolite sand barrier developed first in the margin of the Kangdian old land and formed a sheet of oolite layer which can be traced in most of the Upper Yangtze area with the progradation of the oolite sand barrier toward east, forming a barrier shoal in the shelf, inside of which formed a lagoon and the outside of which was the shallow water shelf, in addition, there were the tidal flat and sabkha in the backshoal (Figure 2a). Figure 1 and Figure 2a show the following paleogeographical elements: (1) the old land, with middle to high mountain topography, leaving mainly the basalt exposed; (2) the alluvial plain - braided stream sediments consisting of purple-red sandstone, conglomerate, sandy mudstone, 0-350 m thick; (3) the

Lecture Notes in Earth Sciences, Vol. 13
T. M. Peryt (Ed.), Evaporite Basins
© Springer-Verlag Berlin Heidelberg 1987

tidal flat - sediments are composed of brown-red banded siltite and mudstone with flaser bedding, 300-400 m thick; (4) the lagoon - brown-red mudstone, marls interbedded with bioclastic limestone containing pelecypods and gastropods, 300-600 m thick; (5) the sabkha - algal-mat limestone interbedded with oolitic, bioclastic limestone, in the top with sabkha dolostone and anhydrite, 200-300 m thick; (6) the oolite sand barrier - massive bioclastic oolite limestone, more than 100 m thick; (7) the shallow water shelf - sediments (300-400 m thick) are mainly thin bedded micritic limestone containing ammonoids.

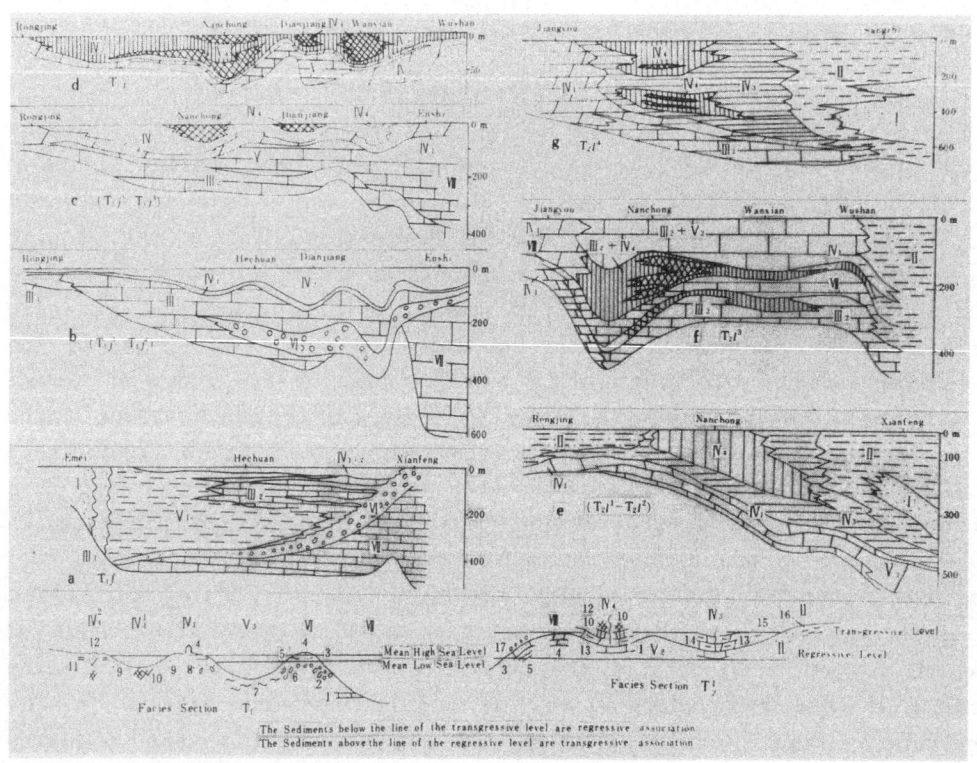

Figure 1.   Paleogeographic map of Feixianguan stage

71

Series	Stage	WEST AND CENTER Member	WEST AND CENTER Formation	EAST Formation	EAST Member
MIDDLE T_2	Ladinian T_2^2	T_2l	Tianjingshan	////////	////////
		T_2l^4			T_2b^4
	Anisian T_2^1	T_2l^3	Leikoupo	Badong	T_2b^3
		T_2l^2			T_2b^2
		T_2l^1			T_2b^1
LOWWER T_1	Olenikian T_1^2	T_1j^5			T_1j^5
		T_1j^4			
		T_1j^3	Jialingjiang		
		T_1j^2			
		T_1j^1			T_1j^1
	Indian T_1^1	T_1f^4		Daye	
		T_1f^1	Feixianguan		T_1d

Table 1. Stratigraphic correlation of Lower and Middle Triassic in the Upper Yangtze area.

During the late Lower Triassic, the sabkha in the platform continued to expand eastward to Wuhan (during T_1j^1- T_1j^2) and to Nanjing (during T_1j^5) due to the shoaling, its distance from east to west being over 1,600 kilometres. In the Upper Yangtze area a series of cycles consisting of lime mud sabkha deposits was formed by the transgression and regression. There are mainly three cycles which are correlated in the area, i.e. T_1j^1- T_1j^2, T_1j^3- T_1j^4 and T_1j^5. During transgression, lime mud was deposited in the tidal flat and lagoon, and calcarenite in the shoal; but during the regression, dolostone and evaporites formed in the tidal flat and the sabkha (Figure 2b,c and d). The paleogeographical elements of late Lower Triassic generally are as follows (from the land to the sea): the old land (it was a plain in geomorphology) - the alluvial plain/eluvial plain - the sabkha platform (playa, sabkha, coastal salt lake) - the tidal flat - the lagoon and shoal - the shallow water shelf - the platform-margin slope - the deep water shelf; see the paleogeographical map of the late stage of late Lower Triassic for a typical example (Figure 3).

At the end of Lower Triassic, the crust rose, so that the platform emerged above the sea level, meanwhile a new old land called Jiangnan was formed in the east. In addition, volcanic activity occurred, a layer of volcanic ash up to 1 m thick was deposited in the whole Upper Yangtze and its neighbouring areas. In some continental depressions, salt playa formed.

The deposition of Middle Triassic is characterized by limestone and evaporite consisting of lime mud - evaporite cycles (i.e. T_2l^1 - T_2l^2, T_2l^3 and T_2l^4), but it mainly shows the basin type of deposition affected by the uplifting and dropping movement of the platform, forming the open sea when the platform was immersed and developing the desiccation basin when it emerged (Figure 2f and Figure 4).

The Indosinian orogeny between the Middle and Late Triassic resulted in the uplifting of the platform and ended the marine deposition in this area.

DISTRIBUTION OF EVAPORITES

The distribution of evaporites in the Upper Yangtze area is shown in Figure 5. Anhydrite covers an area of 500,000 sq.km but halite only occurs in the following thirteen regions: 1 - Nanchong, 2 - Zigong, 3 - Weiyuan, 4 - Chengdu, 5 - Jiangyou, 6 - Daxian, 7 - Dianjiang, 8 - Xuanhan, 9 - Wangcang, 10 - Qijiang, 11 - Wanxian, 12 - Jiannan, 13 - Zhongxian.

Figure 2. Facies section showing depositional environments and models of Early and Middle Triassic in the Upper Yangtze area.
I - alluvial plain, II - shoreside plain, III - tidal flat, IV - sabkha and salt lake: IV_1 - tidal flat, sabkha (dolostone dominated); IV_2 - sabkha salt bank (gypsum dominated); IV_3 - temporary desiccation lagoon (argillaceous dolostone dominated); IV_4 - salt lake (IV_4^1 - coastal salt lake, IV_4^2 - playa). V - lagoon and bay (V_1 - containing red bed, V_2 - lower energy, V_3 - high energy), VI - barrier islands, VII - shallow-water shelf, VIII - platform marginal bank, 1 - micritic limestone, 2 - bioclastic, oolitic limestone, 3 - calcarenite, 4 - dolostone and nodular anhydrite, 5 - algal-mat limestone, 6 - bioclastic limestone, 7 - wormy limestone, 8 - pelletal limestone, 9 - sucrosic, banded anhydrite, 10 - halite, 11 - laminated anhydrite, 12 - polyhalite, 13 - argillaceous dolostone, 14 - shaly limestone, 15 - red sandstone, 16 - red mudstone, 17 - diagenetic dissolution.

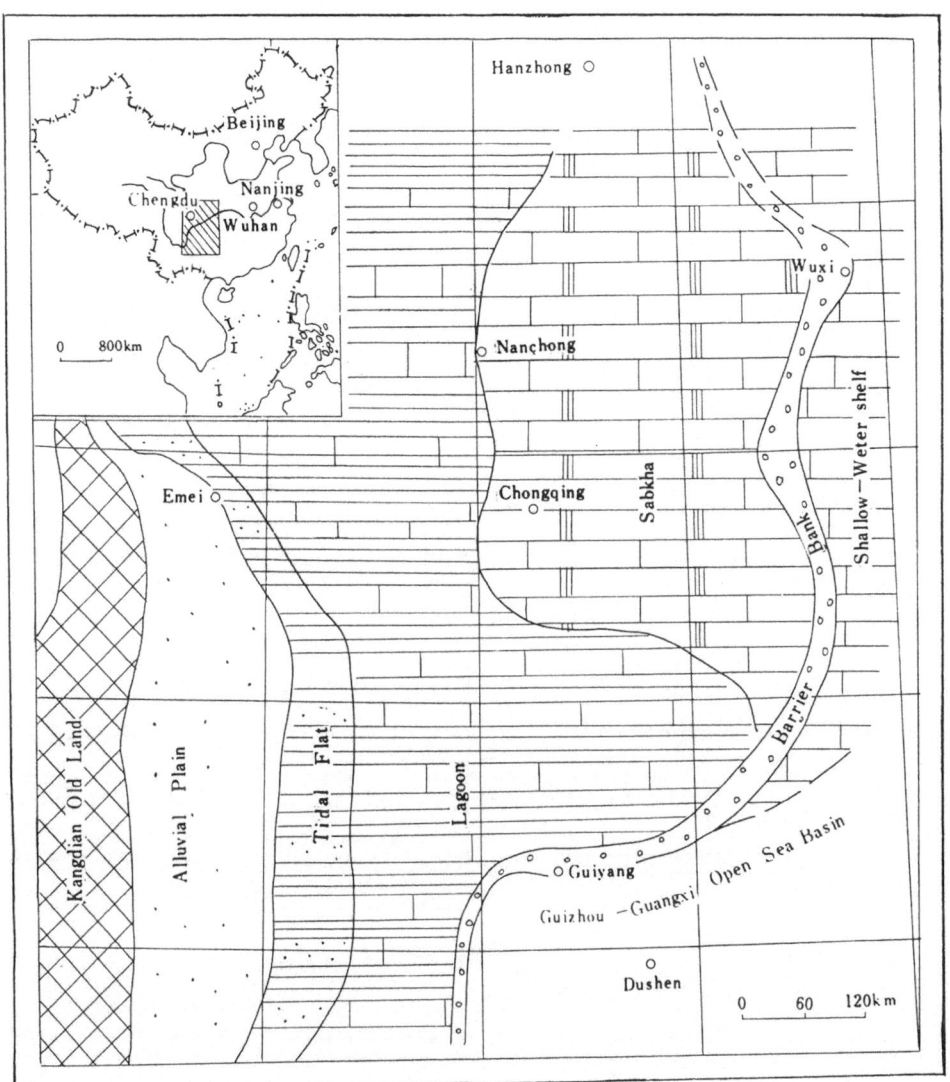

Beijing

Chengdu Nanjing
Wuhan

0   800km

Hanzhong ○

Wuxi ○

Nanchong ○

Emei ○   Chongqing ○   Sabkha   Bank   Shallow—Weter shelf

Kangdian Old Land   Alluvial Plain   Tidal Flat   Lagoon   Barrier

Guiyang ○   Guizhou —Guangxi Open Sea Basin

Dushen ○

0   60   120km

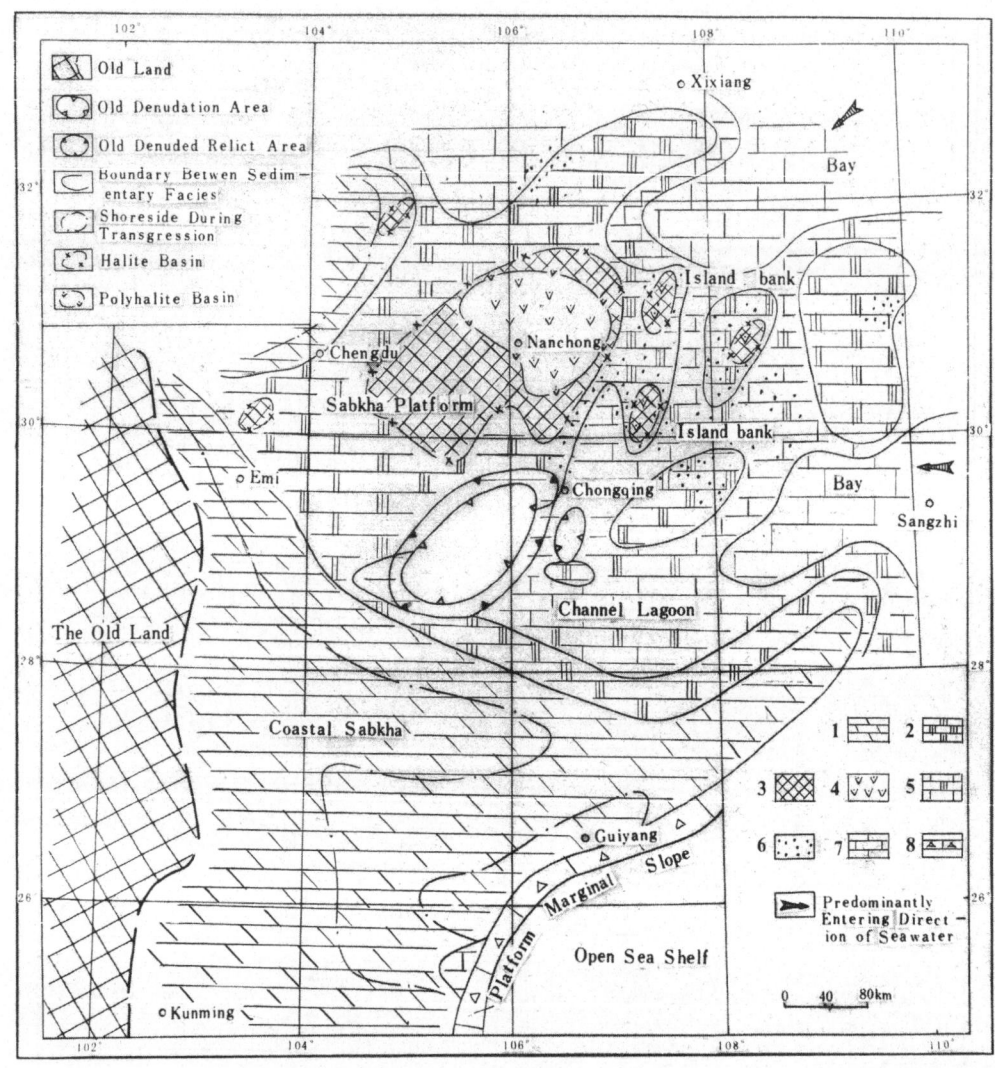

Figure 3. Paleogeographical map of Jialingjiang stage of late Lower Triassic in the Upper Yangtze Platform.
1 - dolostone, in part interbedded with nodular anhydrite or thin bedded limestone, 2 - bedded anhydrite with minor amount of calcarenite, dolostone, in part with halite and polyhalite, 3 - occurrence of halite (the maximum thickness of 60 m), 4 - occurrence of polyhalite in halite, 5 - micritic limestone, with minor amount of dolostone and anhydrite interbeds, 6 - occurrence of oolite, 7 - limestone, with minor amount of dolostone and gypsiferous clay, 8 - laminated limestone, wormy limestone interbedded with slump breccia.

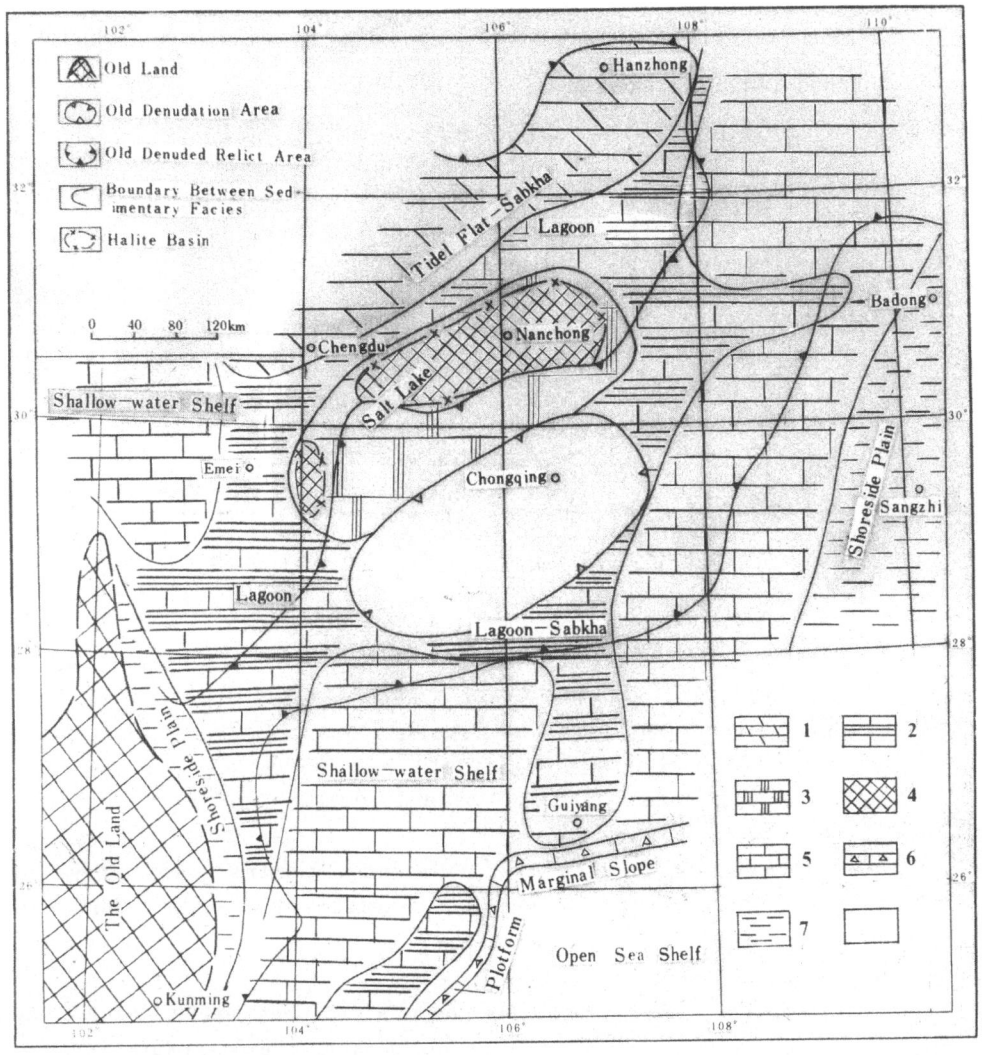

Figure 4. Paleogeographical map of Leikoupo stage of Middle Triassic in the Upper Yangtze Platform.
1 - micritic dolostone, dolarenite with solution pores, and algal-mat dolostone, 2 - shaly marls, micritic limestone, in part with minor amount of micritic dolostone or anhydrite, 3 - marls and limestone, with bedded anhydrite, 4 - marls interbedded with thick-bedded halite, 5 - limestone and nodular limestone with shaly marls, 6 - nodular limestone and slump breccia, 7 - marls and brown-red mudstone.

The region 1, Nanchong, covering an area of about 30,000 sq. km is the largest one among them. Polyhalite was found in region 1,2,4,7,8,and 11. The polyhalite in Nanchong, the region 1, covers an area of 10,000 sq.km.

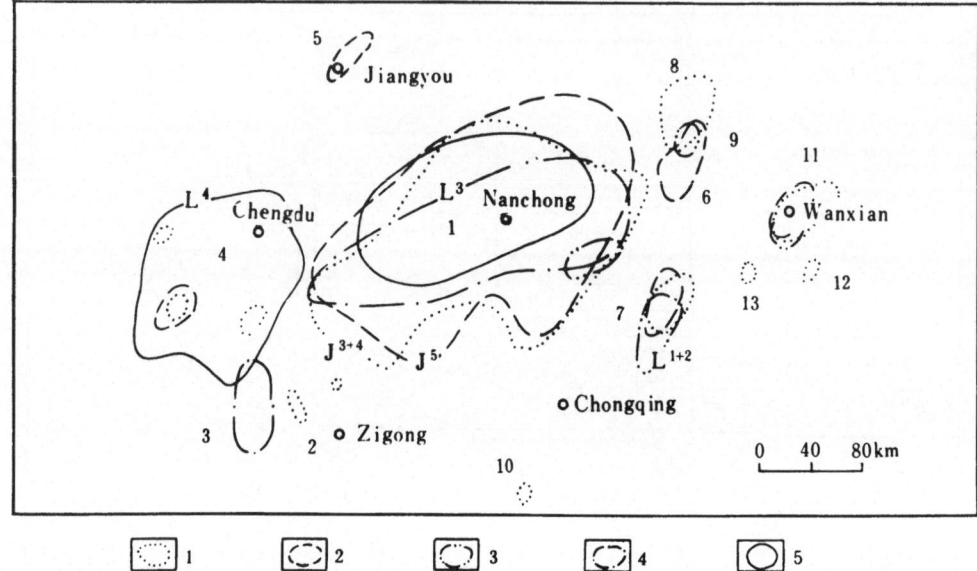

Figure 5. Map showing the distribution of salt bodies of Lower and Middle Triassic in the Upper Yangtze area.

The distribution of evaporites in time shows seven salt-forming stages. All evaporites occur in the upper part of each cycle.

(1) Feixianguan stage: Only anhydrite was found.

(2) Jialingjiang stage 1: Thin bedded halite up to 3 m thick was found in region 1 and 2.

(3) Jialingjiang stage 2: Halite occurs in regions 1,2,4,6,7,8,9,10,11,and 12. In region 1 halite has the maximum thickness of 30 m and the maximum area of 30,000 sq. km. Polyhalite occurs in region 1 and 2.

(4) Jialingjiang stage 3: Halite occurs in regions 1,4,5,6,7, and 11 and has the maximum thickness of 40 m and the maximum area of 30,000 sq. km in region 1. Polyhalite occurs in region 1,6,7,and 11 and covers an area of approximately 40,000 sq. km.

(5) Leikoupo stage 1: Halite occurs in regions 1,6,7,and 11 and has the maximum thickness of 100 m in region 11. In addition, there is polyhalite in regions 1,6, and 11.

(6) Leikoupo stage 2: Halite occurs in regions 1,3, and 12 and has the maximum thickness of 160 m and the maximum area of about 15,000 sq. km in region 1.

(7) Leikoupo stage 3: Halite occurs in regions 1 and 4, but polyhalite only occurs in region 4; in region 4 the maximum thickness of halite is of about 80 m.

ROCK TYPES AND ORIGIN OF EVAPORITES

Evaporites in the Upper Yangtze area mainly consist of anhydrite, halite, polyhalite, and K, Mg-sulfate.

Anhydrite rocks: Residual-primary structures and pseudocrystals of gypsum are preserved, which can be used for interpretation of depositional environments.

(1) Swallow-tail anhydrite rock (Figure 6n): It is uncommon and occurs in T_1J^4, T_1J^5 and T_2J^3. The swallow-tail anhydrite less than 5 cm long was formed in shallow-water environment, presumably in the water less than 5 m deep (Kendall, 1984).

(2) Algal-mat anhydrite rock (Figure 6m): It mainly occurs in T_1J^5 and T_2l^3. The algal-mat consists of micritic carbonates. The anhydrite appears as pseudorhombs or prismatic aggregates less than 0.5 - 1 mm long.

(3) Laminated banded anhydrite rock (Figure 6b,c): The mm-cm lamination consists of argillaceous carbonates (calcite, dolomite amd magnesite) and crystalline anhydrite Some laminae are clear and flat, some are irregular, and with small desiccation cracks which are occasionally common. Crystalline anhydrite is commonly after primary gypsum crystals and its prismatic rhombs are less than 0.5 mm long. In argillaceous carbonates there is smaller microcrystalline or needle anhydrite, some of which were precipitated directly from the brine. The two kinds of crystalline anhydrite are difficult to precipitate from deep water, and for this reason, their depositional environment may be considered as the shallow-water to occasionally subaerial mud flat.

(4) Sucrosic anhydrite rock (Figure 6g,k): It commonly occurs in T_1J^4 and T_2l^4. These anhydrite are of clastic origin. We can see intermittent laminations which give the evidence for the mechanical deposition (Figure 6g). These kinds of anhydrite may be considered as lacustrine coastal sands.

(5) Nodular and enterolithic rock (Figure 6a,d,e and p): The nodule of anhydrite is an aggregate of gypsum monocrystals (Figure 6a) or petal gypsum (Figure 6d) which was formed in the supratidal vadose zone.

The anhydrite rocks (3),(4) and (5) mentioned above are most common.

Halite rocks: They are xenomorphic-unequigranular mosaic in texture and can be subdivided into the following subtypes by structures of the rocks.

78

Figure 6. Rock types, textures and structures of evaporites in the Upper Yangtze area.
a - nodular anhydrite preseving crystal form of monocrystalline gypsum, core is 15 cm long, T_1J^4, b - laminated anhydrite, core 14 cm long, T_2l^1, c - banded anhydrite, core 22 cm long, T_1J^4, d - petal anhydrite, thin section, x 34, crossed polars, e - enterolithic anhydrite, core 10 cm long, T_1J^4, f - floor halite, core 9.5 cm across, g - sucrosic anhydrite, core 16.5 cm long, T_1J^5, h - nodular polyhalite, core 9.2 cm long, T_2l^1.

79

i - halite (dark) polyhalite (light), thin section, x 34, xp., T_2l^1,
j - anhydrite residue in polyhalite, thin section, x 34, xp., T_2l^1,
k - sucrosic anhydrite, thin section, x 34, xp., T_1j^4,
l - monocrystalline gypsum (Gy), showing that the radiating
microcrystalline polyhalite (Po) grew in its margin: Ha - halite, thin
section, X40, m - algal-mat anhydrite, core is 11 cm long, T_2l^3, n -
swallow-tail gypsum, core 10 cm long, o - banded polyhalite, core 25 cm
long, p - nodular anhydrite, core 14 cm long, T_1j^4, q - banded halite,
r - halite with mottled anhydrite, core 19 cm long, T_1j^5.

(1) Floor halite (Figure 6f): It is uncommon and appears as brown grey garnet-like druse, and its crystals are about 0.5 cm in diameter.

(2) Lump anhydrite rock (Figure 6r): This subtype of halite rock makes up 60 - 70 percent of the halite rock. Lumps of anhydrite in halite consist of laminated or banded anhydrite and sucrosic anhydrite and were broken into conglomeratic structure be the recrystallization of halite in diagenesis; it is supposed that the original rock has been halite interbedded with thin-bedded anhydrite.

(3) Banded halite rock (Figure 6q): It consists of white banded halite interbedded with black banded halite containing more small anhydrite nodules. Each group of band is 3 - 11 cm thick, which is considered as the seasonal-depositional cycles in the salt lake.

Polyhalite rocks: They were essentially formed by replacing gypsum or anhydrite. Figure 6j shows the embayed anhydrite residue replaced by polyhalite. According to structures, polyhalite rocks can be subdivided into the following subtypes:

(1) Bedded polyhalite rock: It is massive in structure and fibrous microcrystalline in texture. The replacement remnant of gypsum or anhydrite is common. The original rock possibly was sucrosic anhydrite rock.

(2) Banded polyhalite rock (Figure 6o): The banded polyhalite rock consists of crystalline polyhalite band (fibrous-microcrystalline texture dominated) and argillaceous polyhalite band (chrysanthemum-like-sphaerolitic texture dominated). The origin of the chrysanthemum-like-sphaerolitic texture resulted from the large anhydrite monocrystal replaced by microcrystalline polyhalite. We can observe the sphaerolitic polyhalite growing in the margin of the anhydrite monocrystal and the petal aggregated anhydrite (Figure 6l). The original rock of the subtype of polyhalite rock is predominantly the laminated or banded anhydrite rock.

(3) Nodular polyhalite rock (Figure 6h): The original rock could be the nodular anhydrite rock.

(4) Halite polyhalite rock (Figure 6i): It is characterized be the mottled polyhalite scattering intergrains of halite. In fact, the mottled polyhalite was formed from the mottled anhydrite in the halite, which was replaced by the intercrystalline brine in halite.

SEQUENCES OF EVAPORITES

There are two types of evaporitic-bearing sections (Figure 7). The type I is predominant in the Upper Yangtze area and commonly occurs in the salt-forming stages of $T_1 J^2$, $T_1 J^4$, $T_1 J^5$ and $T_2 1^1$. It can be subdivided into three essential subtypes of sequence considering the features of evaporite rocks and their assemblages.

The sequence I_1 (Figure 7) commonly occurs in $T_1 J^2$, $T_1 J^4$. The halite contains mottled anhydrite. The variation of bromine-chlorine coefficient (Br*1000/Cl) in the halite is approximately between 0.3 - 0.4 (Br: 190 - 240 ppm), maximally more than 0.5 (Br: 305 ppm). The sequence is interpreted as reflecting the evolution from lagoon (unit a and b) to intertidal flat (unit c and d), followed by sabkha, which shows much analogy with the sabkha cycles in Abu Dhabi coastal flats described by Evans (1969) and Butler and others (1982).

The sequence I_2 mainly occurs in $T_1 J^5$ and $T_2 1^1$. Mottled anhydrite halite dominates and is interbedded with banded halite in the middle part. The value of bromine-chlorine coefficient (Br*1000/Cl) in the halite with approximately 0.2 (Br: 110 ppm) in the lower part increases upward abruptly up to more than 0.4 (Br: 220 ppm), maximally 0.6 (Br: 345 ppm), in the top. At the top, anhydrite is interbedded with thin polyhalite halite (unit g), bromine content of which drops to less than 250 ppm. The sequence is interpreted as reflecting the evolution from lagoon (units a and b) to intertidal flat (units c and b) to intertidal flat (units c and d), followed by coastal salt lake (unit e) and sabkha salt flat (units f and g).

Depositional environments of this sequence would be explained in detail as follows.

(a) The environment shown be unit c, the dolarenite, can correspond to an intertidal sand flat.

(b) The environment marked by unit d, the algal-mat anhydrite, is matched with upper intertidal to lower supratidal zone.

(c) The boundary between the units d and e may correspond to mean high sea level to spring-tidal level, where salt pans are located by the opinion of D.R. Stoddart (lectures at the University of Nanjing, China, 1981). The coastal salt lake (unit e) is well matched with the salt pan.

(d) The shelly dolostone in the unit f shows the farthest location affected by spring tide which pushed the shells on to the bank. The shelly dolostone is poorly sorted and contains both the nodular anhydrite formed by vadose action and halite filled in intershell pores by the evaporation of ground water.

(e) The halite with polyhalite in unit g shows its depositional position where seawater was not directly supplied and was only recharged by phreatic water deriving from both continental and the sea water considering a tendency towards decrease of bromine content.

The typical features of sabkha in the sequence I_2 are shown in the units f and g. The unit e represents the depositional environment of supratidal salt pans which are well included into the sabkha by the large geomorphologic landscape. In fact, both salt pan and sabkha deposits exist in the same stratigraphic unit.

The sequence I_3 mainly occurs in T_1J^4, T_1J^5, T_2I^1 and T_2I^4, and it essentially is a series of laminated magnesite anhydrite interbedded with many layers of thin bedded polyhalite or/and halite, some of which contain kieserite and langbeinite. Also there is glauberite layer occurring in the upper part of this sequence (Figure 7). This subtype of sequence commonly occurs above the sequence I_1 and I_2 mentioned above. The values of carbon and oxygen stable isotopes in carbonates of the studied area suggest that this sequence was formed in a continental environment (Table 2).

Samples (the same age)	$\delta^{18}O$	$\delta^{13}C$
Sequence I_3 carbonates	-13.75	-19.71
Sabkha carbonates	- 1.84	- 5.23
Open-sea carbonates	- 4.62	+ 2.23

Table 2. Average $\delta^{18}O$ and $\delta^{13}C$ values.

The type II of evaporite section which is characterized by lack of dolostones is uncommon (Figure 7). It occurs only in the salt-forming stage of T_2I^3. The micritic limestone (unit a) contains ammonoids. Mottled anhydrite halite is interbedded with banded halite (unit d), the bromine-chlorine coefficient in the halite is commonly less than 0.1, between 0.05 and 0.08 (Br: 25 - 45 ppm, maximally 150 ppm), with small variation ranges. This type of sequence seems to reflect the evolution from the open sea (unit a) to lagoon (unit b), followed by salt lake (units c and d) and sabkha (unit e). It must be pointed out that the tidal flat or sabkha dolostone under the evaporite is absent from the type II of evaporite section as compared with the type I.

DEPOSITIONAL MODELS OF EVAPORITES

The deposition of Triassic evaporites in the Upper Yangtze area can be ascribed to two types of depositional models according to the section and characteristics of the sequence. The evaporites with characteristics of the type I are included in the platform sabkha model, and the evaporites with characteristics of the type II are included in the inner-platform lagoon model.

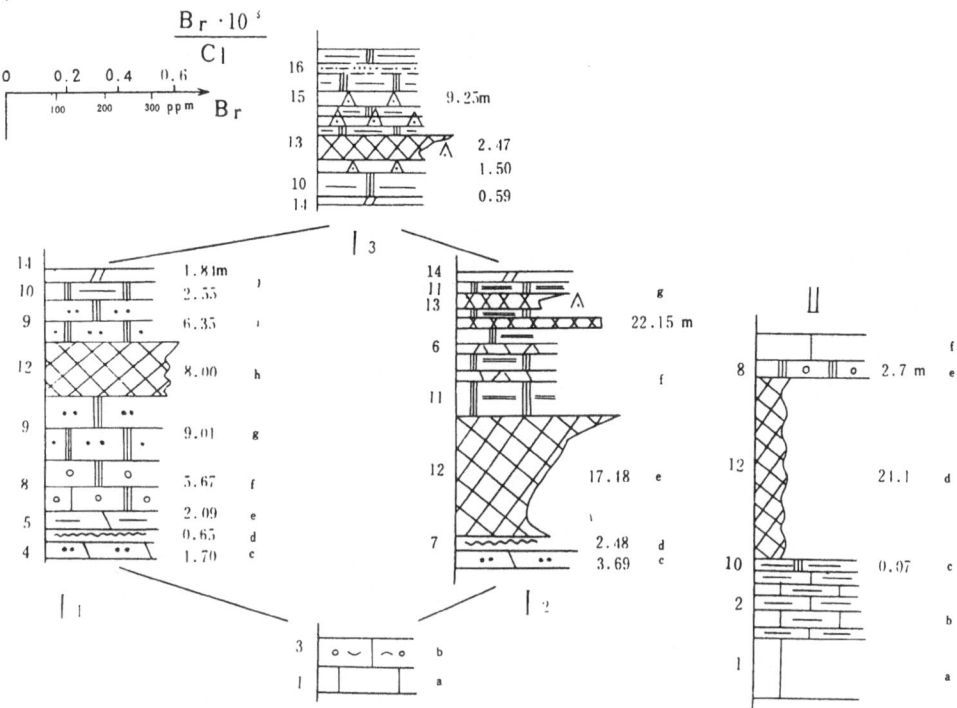

Figure 7. Types of sequences of Triassic evaporites of the Upper Yangtze area.
1 - micritric limestone, 2 - shaly marls, 3 - bioclastic oolitic limestone, 4 - dolarenite, 5 - argillaceous dolostone, 6 - shelly dolostone, 7 - algal-mat anhydrite, 8 - nodular anhydrite, 9 - sucrosic anhydrite, 11 - banded anhydrite, 12 - halite, 13 - halite with polyhalite, 14 - mudstone with magnesite, 15 - polyhalite, 16- volcanic ash, a-j - units of the section from bottom to top.

Platform sabkha model. Evaporites occurred in the supratidal flat which was formed by progradation of the coast. This model essentially bears analogies with the recent sabkhas of the Persian Gulf but the topography of the platform was not such flat as in Abu Dhabi. There were some depressions in the margin and inner part of the platform. The marginal depressions were coastal lakes formed by the bank which confined the channel of the lagoon; the inner depressions were formed by diversified subsidence in the sabkha platform.

The evolution of the sabkha shown as the section sequence I (Figure 7) can be divided into three stages (Figure 8).

(a) Coastal salt lake (salt pan) stage. The salt lake was a shallow-water salt pan in the progradational coast closed by the sea

84

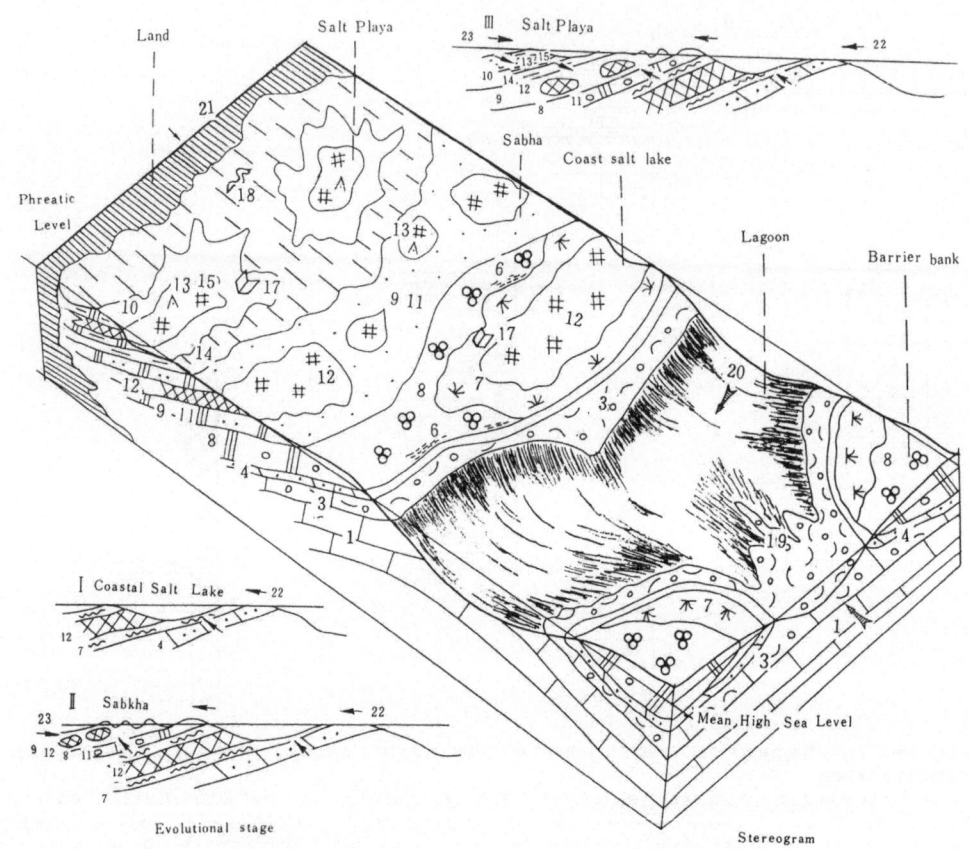

Figure 8. Model for deposition of Triassic sabkha platform evaporites in the Upper Yangtze area.
1-16 - see Figure 7 for explanation, 17 - swallow-tail gypsum, 18 - enterolithic gypsum, 19 - tidal bar, 20 - entering seawater, 21 - main direction of continental water supply, 22 - recharge by seawater, 23 - recharge by continental water.

beach. The altitude of the salt pan was controlled by the mean high sea level. The salt pan was supplied with sea water by flood flow. Evaporation from both brine surface and phreatic water existed when evaporites were precipitating. In this stage, a dominant sediment was halite.

(b) Sabkha stage. Evaporites were here formed in three ways: First, when the sabkha was flooded during spring tide, laminated gypsum and micritic carbonates (dolomites dominated) would be deposited, this is main way to precipitate gypsum. Second, in some depressions, halite precipitated in intercrystalline pores of gypsum by phreatic evaporation because of higher phreatic table, even in case of groundwater recharge, a temporary salt lake would form, were the thin halite layer with high conent of bromine precipitated. Third, in some lowland where was near land and flood recharge did not replenish, while halite was precipitated, also the .halite with polyhalite occurred. The polyhalite resulted from the continental groundwater supply carrying Ca and SO ions to form gypsum, then Ca ion of which was replaced by K^+ and Mg^{2+} ions.

(c) Playa lake. The coastal progradation and the rise of the platform made the coastal sabkha entirely lost contact with seawater to develop the playa where was a new depression formed by diversifield subsidence in the original sabkha plain. The playa was supplied by surface water and groundwater deriving from continent and rainwater which dissolved the salt in original sabkha, so that the composition of the brine in the playa was characterized by lower concentration and being rich in K and Mg ions. This resulted in a sequence in which halite is uncommon and laminated anhydrite and bedded polyhalite dominate.

As a result of the development of these three stages, corresponding three lithofacies zones occurred laterally in the progradational coast of the platform.

The progradation of the sabkha platform developed in the regressive background. As shown in Figure 3 and Figure 2d, it first formed the bay along the eastern margin of the platform and the earliest coastal sabkha. Meanwhile some island banks (offshore banks) which were barriers to separate the lagoon from the bay began to form in the eastern margin of the platform. The progradation of the sabkha platform essentially expanded towards the lagoon from the shoreline, meanwhile there was the progradation towards the lagoon from the large offshore banks (cf. Figure 2d). As the sea level fell, the sabkha platform turned to the playa, and the coastal sabkha was led to progradation towards the lagoon. The progradation first took place in some island banks (e.g. in Dianjiang), then towards the shallow-water shelf from the lagoon. It must be pointed out that the progradation process developed so rapidly, that the sabkha platform expanded to the Middle and Lower Yangtze area from the Upper Yangtze area, forming a sabkha plain covering an area of 6,000,000 sq. km. In addition, as the groundwater table fell, the playa would move into the original lagoon depression, for the platform to evolve into eluvial plain.

It seems that halite was deposited only in the sabkha platform and islands banks, but there was no halite deposition in lagoon in the shallow-water shelf. The saline mineral zonation in each salt lake which formed in the sabkha platform and island banks appears as a bulls-eye pattern, so it shows the characteristic of the multicentral bulls-eye pattern in the whole area.

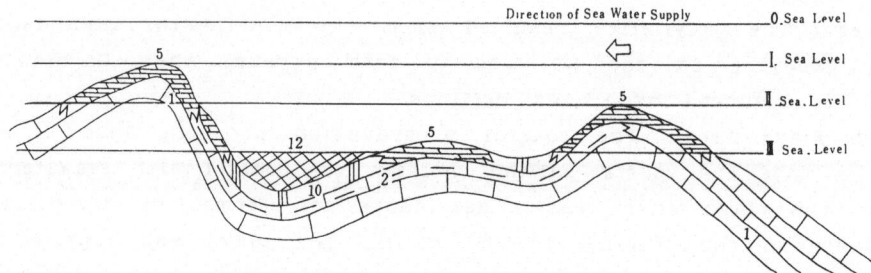

Figure 9. Desiccation-lagoon depositional model of Triassic evaporites in the Upper Yangtze area (see Figure 7 for explanation).

Desiccation-lagoon model in the platform. It was the salt lake separated from the lagoon in the platform by regression, not be the progradation of coast. The bottom of the salt lake was not exposed, but the marginal beach of it would be exposed, where dolomitization and dissolution occurred. The depositional model is shown in Figure 9.

As sea level fell to the level I from the level O, the margin of thr platform was exposed, but in the inner of the platform the lagoon formed. When the sea level fell to the level II, the lagoon was broken up into some separated basins which evolved into salt lakes, where micritic dolostone and gypsum were formed in the beach. With the further fall of sea level, the salt lake shrank and precipitated halite and gypsum in the centre and gypsum and dolostone on the beach. The dissolution occurring in the lake margin resulted in pore formation in the dolostone and the calcitization of gypsum.

The salt lake was fed temporarily by flood seawater. The content of bromine in the halite indicates the halite underwent dissolution-redeposition. After the salt lake was entirely filled with halite, it would be turned into the coastal sabkha. Because the seawater supply was periodic, the saline mineral zonation in a salt lake still appears as the bulls-eye pattern, its centre is near the position of the seawater supply (in the north-east of the salt lake).

CONCLUSIONS

The evaporites discussed above formed in the inner shelf of the platform, and evaporite basins were restricted by time-parallel strata. The general geological setting and the development of evaporites suggest that the basins were flat and shallow, and were formed in two ways: one is to develop in the progradational-coastal sabkha were seaside depressions formed by subsidence, the other, relic lake formed in inner shelf depressions by rapid regression. The two types of models are different in evaporite association and its sequence. The differences, however, would gradually disappear with the evolution from the salt lake to playa by gradual separating from seawater.

Saline mineral zonation of both models appears as bulls-eye pattern zoning for an evaporite basin, and multicentral bulls-eye pattern for the whole platform.

REFERENCES

Bathurst R.G.C., 1971 - Carbonate Sediments and Their Diagenesis. Elsevier, Amsterdam, London, New York, 620 pp.

Butler G.P., Harris P.M. & Kendall C.G.St.C., 1982 - Recent evaporites from the Abu Dhabi coastal flats. SEPM Core Workshop 3, 33 - 64.

Evans G. & Bush P.R., 1969 - Some sedimentological and oceanographic observations on a Persian Gulf lagoon. UNESCO Conference on Coast Lagoon, Mexico City, 1967, 150 - 170.

Ginsburg R.N.(ed.), 1975 - Tidal Deposits: a Casebook of Recent Examples and Fossil Counterparts. Springer-Verlag, Berlin, 428 pp.

Handford C.R., 1981 - Coastal sabkha and saltpan deposition of the Lower Clear Fork Formation (Permian), Texas. J. Sed. Petrol., 51, 761 - 778.

Hsu K.J., 1972 - Origin of saline giants: a critical review after the discovery of the Mediterranean evaporite. Earth-Science Reviews, 8, 371 - 396.

Hsu K.J., He Qixiang, Wu Yinglin & Zhu Zhongfa, 1983 - Carbon - and oxygen - stable isotope geochemistry of the carbonates above and below the synchronous marker 'mung bean' rock between the Early and Middle Triassic in the southwest China. Bull. Chengdu Inst. Geol. Min. Res., no. 4, 1 - 12.

Illing L. V., Wells A.G. & Taylor J.C.M., 1965 - Penecontemporary dolomite in Persian Gulf. SEPM Spec. Publ. 13, 89 - 111.

Kendall A.C., 1984 - Evaporites. In: R.G. Walker (ed.), Facies Models. 259 - 296.

Kerr S.D. & Thomson A., 1963 - Origin of nodular and bedded anhydrite in Permian shelf sediments, Texas and New Mexico. Bull. Amer. Assoc. Petrol. Geologists, 47, 1727 - 1732.

Kinsman D.J.J., 1969 - Models of formation, sedimentary association, and diagnostic features of shallow-water and supratidal evaporites. Bull. Amer. Assoc. Petrol. Geologists, 53, 830 - 840.

Reading H.G.   (ed.), 1978 - Sedimentary Environments and Facies. Oxford London, Blackwell Scientific Publication.

Richter-Bernburg G., (ed.), 1972 - Geology of Saline Deposits. UNESCO, Paris.

Schmaltz R.F., 1969 - Deep-water evaporite deposition - a genetic model.Bull. Amer. Assoc. Petrol. Geologists, 53, 798 - 823.

Shinn E.A., Lloyd R.M. & Ginsburg R.N., 1969 - Anatomy of a modern carbonate tidal flat, Andros Island, Bahamas. J. Sed. Petrol.,39, 1202 - 1228.

Wilson J.L., 1975 - Carbonate Facies in Geologic History.   Springer-Verlag, Berlin, Heidelberg, New York. 471 pp.

Zeng Yunfu, Lee Nan Hao & Huang Yang-Zohn, 1983 - Sedimentary characteristics of oolitic carbonates from the Jialing-Jiang Formation [Lower Triassic (T j)], South Sichuan Basin, China. In: T.M.   Peryt (ed.), Coated Grains, Springer-Verlag, Berlin-Heidelberg, 176   -   187.

MIDDLE MUSCHELKALK EVAPORITIC DEPOSITS IN EASTERN PARIS BASIN

D.Geisler-Cussey
Laboratoire de Géodynamique des Bassins Sédimentaries
Faculté des Sciences, Université de Pau
Avenue Philippon , 64000 Pau
France

INTRODUCTION

Middle Muschelkalk evaporitic deposits in Eastern Paris Basin are in the small basin of Lorraine part of the large German Basin. The main interest in this area is that it is located on the basin margin, close to the shore, were continental influences will be expressed in the sedimentary record. Recent drilling, with continuous coring, allows analysis of the vertical distribution of facies and therefore aids in the understanding of the depositional mechanisms with a dynamic approach to the study of the basin.

PALEOGEOGRAPHIC AND STRATIGRAPHIC FRAMEWORK

PALEOGEOGRAPHIC DATA

Many synthetic paleogeographic maps project the extension of the Muschelkalk series together with location of the evaporitic deposits in Western Europe (Ricour, 1962 ; Wurster, 1964 ; Wild, 1968 ; Trusheim, 1971 ; Richter-Bernburg, 1972). Drilling for oil in the North Sea has brought additional information in the subsea areas (Ziegler, 1982).

The paleogeographic maps of the saliferous Muschelkalk in Western Europe show the continuity of salt as a main basin with a west to east elongation from the North Sea to northern Germany and a narrow channel extending from NNE to SSW down to Switzerland (Figure 1). The basin of Lorraine is just an appendix to the main German Basin

Lecture Notes in Earth Sciences, Vol. 13
T.M. Peryt (Ed.), Evaporite Basins
© Springer-Verlag Berlin Heidelberg 1987

and was close to land areas during deposition: Rhenan and Brabant Massifs represent the Ardennian Continent. Sea water inflow is from the north east.

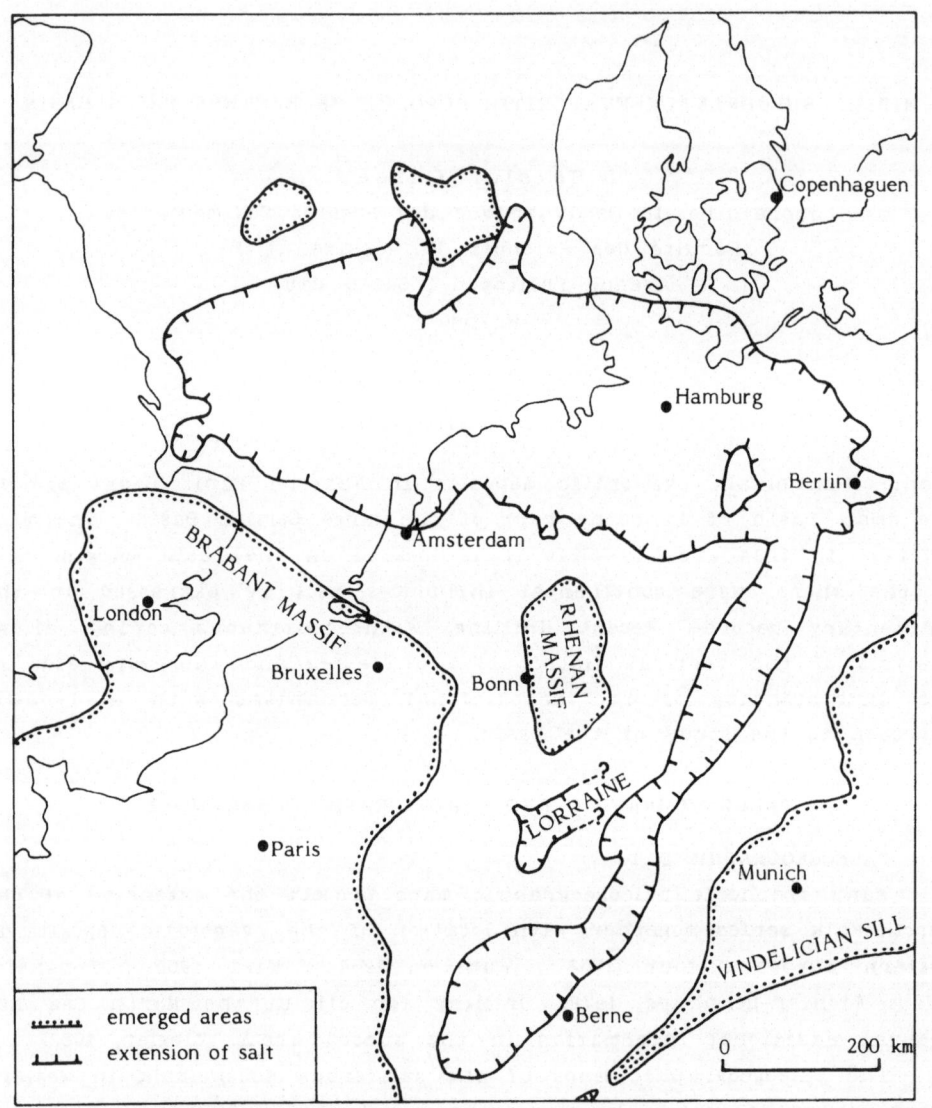

Figure 1. Paleogeographic map of the saliferous Muschelkalk in Western Europe (from Ziegler, 1982).

The Middle Muschelkalk basin of Lorrain contains a central saliferous unit, 75 km wide from north to south, beetwen Faulquemont and Baccarat and 50 km across the eastern limit of outcrops and its western edge (Fourmentraux et al., 1959). The saliferous zone is surrounded with an outer rim of sulfate and argillaceous deposits extending 150 km towards the west to the region of Sainte-Menehould and Saint-Dizier (Figure 2). The location of the basin corresponds to the present Sarreguemines syncline (Le Roux, 1971) and has on its western rim, two slight extensions to the NW and the SW.

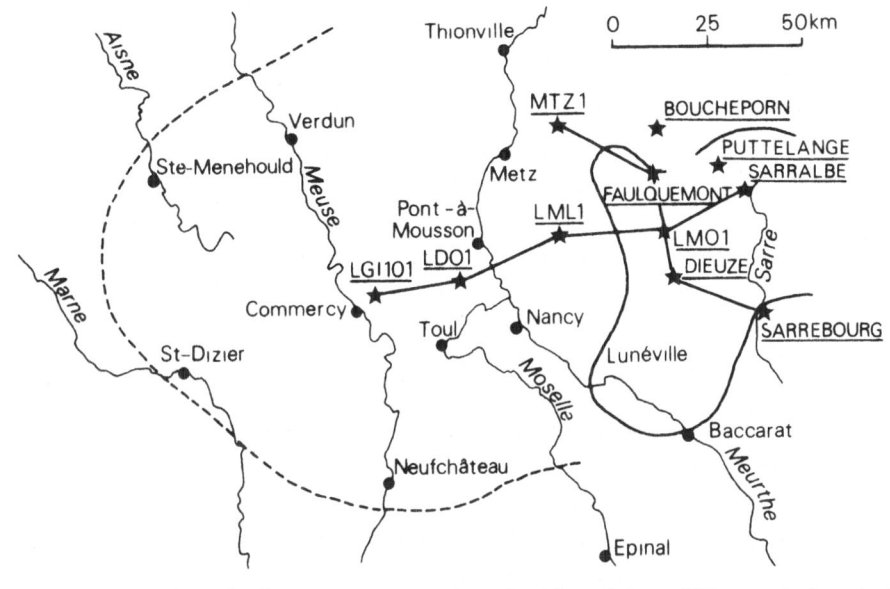

———— extension of salt   – – – – extension of sulfate-rich argillites   ★ bore holes

Figure 2.Extension of the Middle Muschelkalk deposits in eastern Paris Basin (from Maget & Maiaux, 1980).

STRATIGRAPHIC DATA

Middle Muschenkalk of Lorraine (110 to 80 m thick) is stratigraphically divided into three lithologic units, according to their prevailing color, from the bottom up: Red Beds, Grey Beds, and White Beds Formations. The Ca-sulfate and halite evaporitic levels are restricted to the Grey Beds Formation in which they are interbedded with grey colored clays; the thickness of the halite varies according to their degree of dissolution (50 to 70 m , or less). The Red Beds Formation (25 m average thickness) is essentially argillaceous, red and green, bright colored, with some sulfate levels, particularly close to the bottom and top. Finally, the White Beds Formation (5 to 10 m average thickness) corresponds to beige colored dolomites and argillaceous dolomites. The transition from the Grey Beds Formation to the White Beds Formation is progressive with development of dolomite towards the top; therefore it is difficult to precisely locate the contact between both units on lithologic sections.

On the other hand the lithostratigraphic limits of Middle
Muschelkalk show a sharp contrast: at the bottom, the clays of the Red
Beds Formation are directly underlain by the dolomites of the Lower
Muschelkalk Myophoria level, and at the top the dolomites of the
White Beds Formation pass without transition to the limestones of the
Upper Muschelkalk series.

It is worth noting that these series become more clearly
detrital approaching the Ardennian shore (van Wervecke, 1916).

GEOMETRY OF EVAPORITIC BEDS

Drilling through the Middle Muschelkalk series of Lorraine was
carried out prospecting for various material: coal, hydrocarbon, salt,
gypsum, anhydrite and water. Except for the oil drilling, there were no
electric logs available; therefore the study is essentially based on
lithologic logs.

Lateral correlations take utilize the top of Middle Muschelkalk as
a the horizontal reference layer,because the contact between dolomites
of the White Beds Formation and the Upper Muschelkalk limestone is easy
to locate on sections. Distribution and extension of halite and Ca-
sulfate beds is followed on cross-sections through basin.

WEST TO EAST CROSS-SECTION

A longitudinal section allows the Grey Beds Formation to be
differenciated into three evaporitic units: anhydritic at the bottom
and the top,and saliferous in between (Figure 3). The anhydrite levels
are followed with minor thickness changes across the section. The lower
unit consists of a single anhydrite bed, while the upper one is
composed of two beds, except on the west side of the basin, where only
one anhydrite bed remains. On the other hand, the middle salifeours
unit is only observed on the eastern half of the section and shows a
progressive westward thinning; it finally disappears completely to the
west of Moncheux drillsite (L ML1). There is also a noticable
thickening of the series where the saliferous unit develops and a
thinner section in the area of Domevre drilling (L DO 1).

All these factors taken together indicate that the zone of
greatest subsidence was located along the axis of present Sarreguemines
syncline, where the saliferous deposits show their maximum thickness.
On the other hand, the Domevre drilling (L DO 1), having the thinner
section, corresponds to an area of lesser subsidence situated on
the present Sarro-Lorrain anticline. Therefore it is obvious that the
main NE-SW Hercynian structures of Eastern Paris Basin basement

93

exerted an important effect on sedimentation rate at Middle Muschelkalk time.

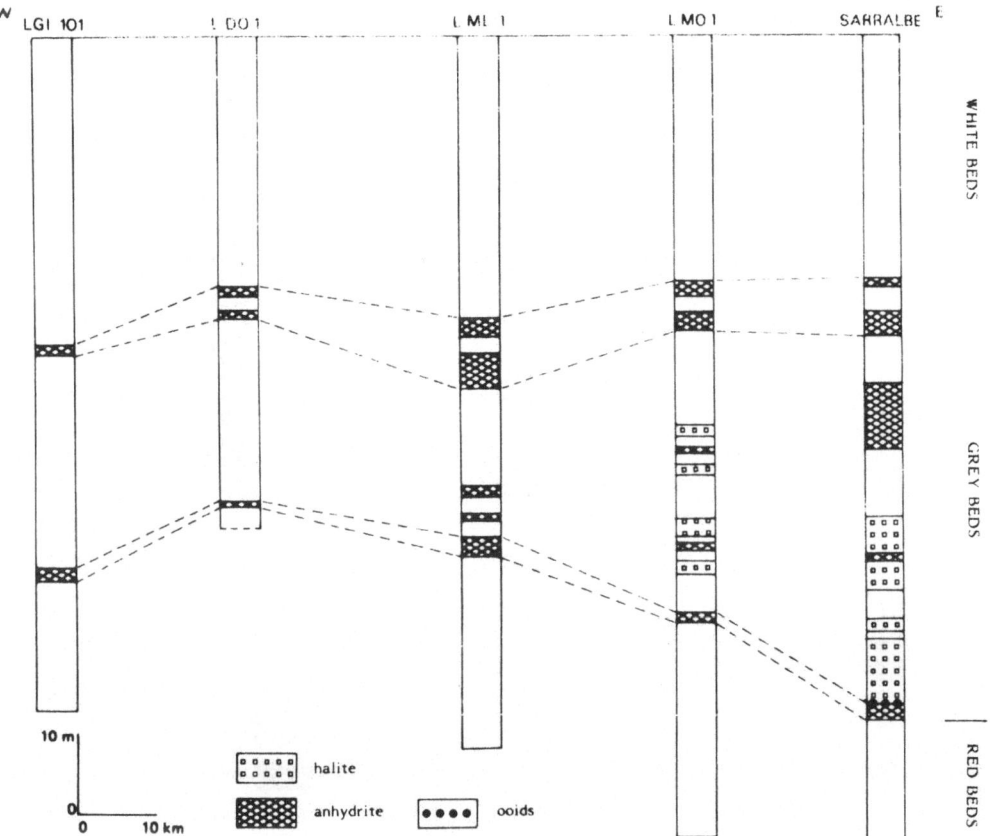

Figure 3. West to east cross-section through the Middle Muschelkalk Basin of Lorraine(after Geisler, 1982b). See location of drillings on Figure 2.

TRANSVERSE CROSS-SECTION NORTH WEST TO SOUTH EAST

A transverse section through the basin shows the same three evaporitic units as on the previous longitudinal section (Figure 4). In the some way the lower and upper anhydrite units are laterally continous, but display important thickness changes. The lower unit is actually very thick in Sarrebourg (15 m) and thins down progressively to the NW towards Faulquemont (3 m). A decimetric oolitic layer, identified at the top of this lower unit in Sarrebourg, Faulquemont and also in the area of Sarralbe (Laugier, 1959), justifies this lateral correlation.

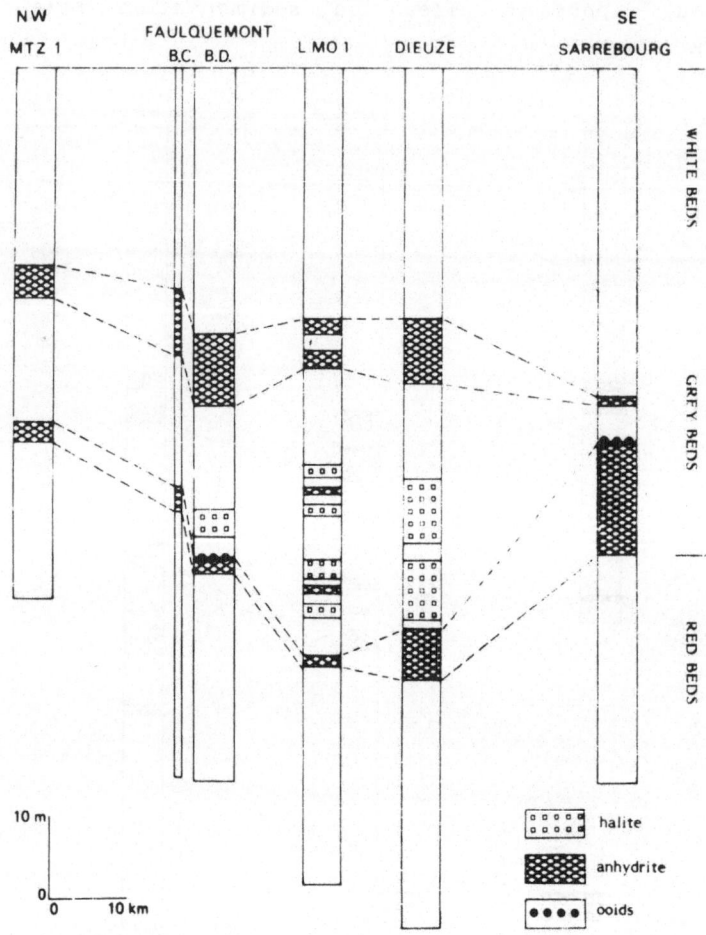

Figure 4. Transverse cross-section north west to south east throught the Middle Muschelkalk Basin of Lorraine (after Geisler,1982b). See location of drillings of Figure 2.

The middle saliferous unit is restricted to the central part of the traverse and is abruptly interrupted on both sides. Here it also corresponds to the greatest thickness of Middle Muschelkalk series and lies in to the area of main subsidence. This distribution of the evaporitic units is in best agreement with the well known arrangement in many evaporitic basins, where saliferous deposits are generally located in the central part, within the area of greatest rate of subsidence.

In general, the lowest anhydrite unit is thicker on the eastern basin edge, while saliferous deposits develop only in the central area.

Because subsidence is marked by the greatest thickness, the Middle Muschelkalk basin of Lorraine shows progressive lateral shifting of subsidence from east to west. It becomes evident at a very early stage of the differenciation of the Paris Basin, which becomes more obvious during deposition of the Keuper salt (Marchal, 1983).

PALYNOLOGICAL DATA

No systematic palynological study of Middle Muschelkalk series has been undertaken until recently (Adloff, Doubinger & Geisler, 1982). Therefore it is of great interest to identify the microfloral associations characteristic of this series, just as their variations depend on different facies and the location of the sections within the basin.

Bore holes, two kilometers apart, located a few kilometres west of Sarrebourg and going right through the entire Middle Muschelkalk, constitute helpful reference series. Two other sections on the northern edge of the basin: Faulqemont and Boucheporn, yield some additional information on palynological content in salt and on influence of the proximity of the Ardennian shore.

PALYNOLOGICAL STUDY AND STRATIGRAPHIC INTERPRETATION

The Red Beds Formations contain a well preserved microflora in which 49 species of spores and pollens could be identified. Among them *Illinites chitonoides*, *Sulcosaccispora minuta*, *Jugasporites renalis* and *Tsugaepollenites oriens* are typical of Middle Muschelkalk. The significant amount of *Triadispora staplini*, the presence of *Illinites chitonoides* and *Hexasaccites muelleri* (syn. *Stellopolenites thiergartii*) and the lack of *Ovalipollis pseudoalatus* permits attribution of an Anisian age, probably Upper Anisian, to the palynological association.

There in an absence of microflora in the White Beds Formation, and a similar association in observed at the bottom of Upper Muschelkalk limestones. Acritarches are very abundant, but pollens and spores are badly preserved.

QUANTITATIVE ESTIMATIONS AND PALEOGEOGRAPHIC IMPLICATIONS

Depending on the manner of dispersal and the probable ecology of mother plants, three groups are distinguished: spores, bisaccate pollen and other pollen. Spores are carried by fresh water out to the sea where they settle more or less close to the shore; therefore their relative abundance is related to connection with restricted transport mechanisms. On the other hand, bisaccate pollens may be transported by wind over long distances before deposition. Quantitative estimates indicate a great abundance of bisaccate pollen and a very small set of spores (5 %). In the main, the palynological association is homogenous in the Red Beds and Grey Beds Formation.

Some differences appear on the northern edge of the basin where the qantity of spores reaches 21% in the salt of Faulqemont and 18% at the top of Boucheporn section. These values indicate a fresh water supply carrying the spores from the Ardennian continent to the northern rim of the Middle Muschelkalk of eastern Paris Basin.

The disappearance of microflora in the White Beds Formation seems to be related to a modification of preservation conditions of organic matter in a more oxidizing environment rather then to an original absence. This basic assumption permits the persistance of the same vegetation on surrounding continental areas and therefore preservation of the same major climate during all Middle Muschelkalk. Reappearance of a similar palynological association at the bottom of Upper Muschelkalk confirms this hypothesis.

However the sharp contact between dolomites at the top of the White Beds Formation and limestones at the bottom of Upper Muschelkalk is an expression of the Upper Muschelkalk transgression, in which an open marine environment is established. This fact is palynologically well expressed by the sudden appearance of Acritarches, which were absent during all Middle Muschelkalk and indicate open marine conditions.

SEDIMENTOLOGICAL DATA

Description, genesis and distribution of the various facies allow us to develop a good idea about depositional environments and mechanisms in the basin. The reference bore holes of Sarrebourg area provide a typical section for the Middle Muschelkalk series (Figure 5). Only the salt is lacking here, but it has been observed in a drillsite in the Faulqemont area.

FACIES DESCRIPTION AND GENESIS

Detrital terrigenous facies

Mainly composed of clay minerals, they are present throughout the entire series. Clay mineral associations and chemistry will be studied in greater detail later in this paper. However, these argillites show variable colors and carbonate contents (Figure 5).

97

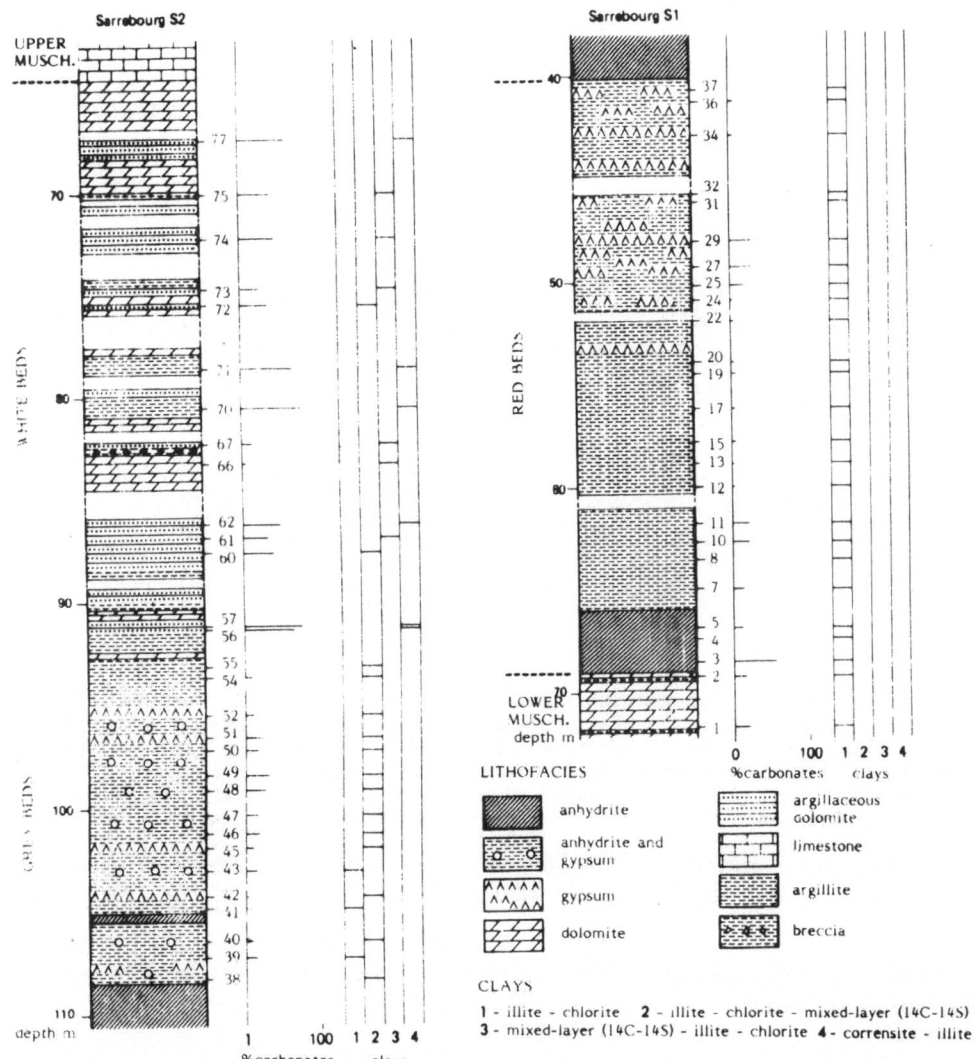

Figure 5. Lithologic logs of the Sarrebourg bore holes, location of analysed clay samples, carbonate contents and distribution of clay mineral associations (after Geisler, 1982b).

In the Red Beds Formation, the argillites are red and green, bright colored, with carbonate contents are generally under 10 %, rarely reaching 30 %.

In the Grey Beds Formation, as indicated by their name, argillites are grey colored, sometimes very dark, with a few brightly colored layers, such as the "Guillaume marker" which is located at the top of the zone of possible salt occurrence. The average carbonate contents remain low, between 10 and 20 %, as in the Red Beds Formation.

In the White Beds Formation, the argillites are light colored, white to beige, and their dolomicritic carbonate content is much higher, between 30 to 70 %. It is also worth noting the occurrence of thin laminated argillite levels in this unit with alternating millimetric laminae of green argillites and white argillaceous dolomites, representing primary depositional sequences (Figure 6).

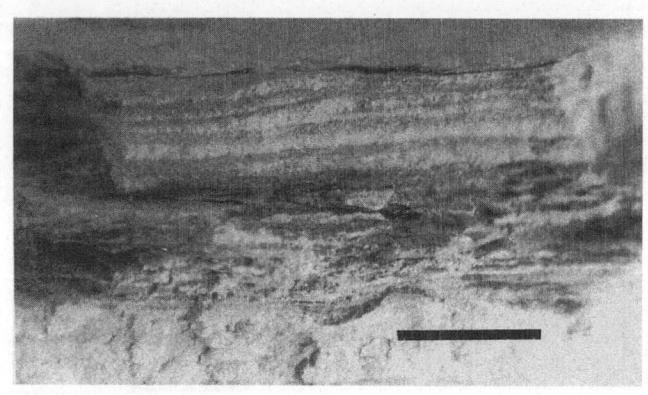

Figure 6. Millimetric alternations of green argillites (dark on the picture) and white argillaceous dolomites, representing primary depositional sequences. On the lower part note the sharp contact beetwen a white dolomite layer topped with green argillites, figuring respectively the top of a lower decimetric sequence and the bottom of the next one. Sarrebourg, White Beds Formation (bar = 5 mm).

Some silt- and sandstone levels interfinger between the argillites at the bottom of the Red Beds Formation in the Faulquemont area. The coarser composition of this detrital terrigenous facies is even more obvious further to the north, towards the Ardennian shore (van Wervecke, 1916).

Oolitic facies

This facies is only encountered at the top of the lower anhydrite unit in the Grey Beds Formation and represents a stratigraphic marker of decimetric thickness. Macroscopic observation shows white millimetric to submillimetric rounded elements, contrasting strongly with the dark background of the rock. On thin section, the oolitic structure is obvious, displaying ooids with a radial pattern for the cortex and nuclei of various kinds, sometimes gastropod shells. The ooids are cemented with poikilotopic gypsum (Figure 7). They represent an episode of carbonate sedimentation, included in Ca-sulfate during an early stage of diagenesis.

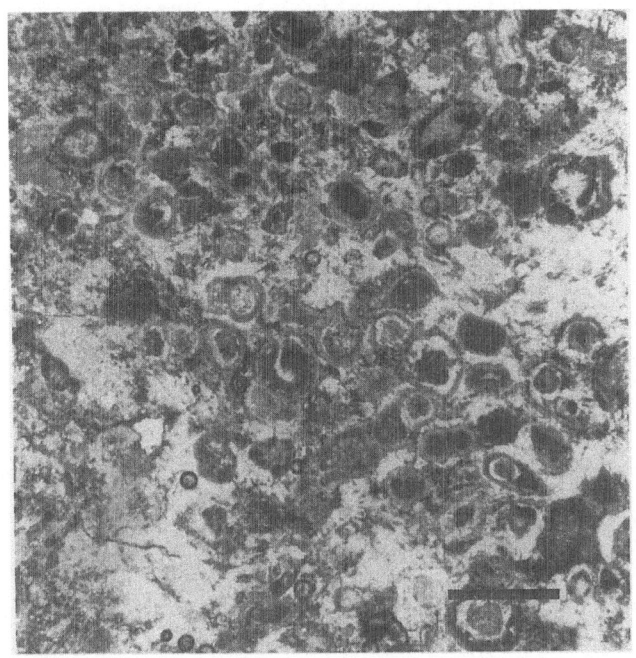

Figure 7. Photomicrograph of carbonate ooids displaying a radial
pattern for the cortex and various kinds of nuclei, comented with
poikilotopic gypsum. Sarrebourg, Grey Beds Foramtion (plane light, bar
= 2 mm).

Dolomite   facies

The dolomites occur in the White Beds Formation and display
various macroscopic patterns, but are mostly dolomicritic, generally
bedded and white to beige colored. The bedding is more or less thin:
centimetric, if related to detrital silt layers or millimetric,
looking like thin algal laminations. Centimetric or even thicker beds
of coarse, lumpy or fine dolomicrite may alternate. Some layers show
intraformational breccias, probably originating from the breaking up of
desiccation chips (Figure 8).

Macroscopic observation shows a very fine dolomicritic matrix
interlayered with generally graded bedded layers, which make up primary
depositional sequences. Various compositions occur in the detrital
layers: silt on gully surfaces, alone or together with black pebbels
(Figure 9), micritic pellets with some Miliolideae (Figure 10) or
micritized pelecypod shells with some ostracods (Figure 11). Genesis of
such deposits seems related to deposition of a very fine dolomitic mud,
probably of syngenetic origin periodicaly interrupted by currents

carrying various detrital particles. Presence of Miliolideae, ostracod and pelecypod shells clearly suggest marine influences.

Figure 8. Light beige dolomite showing in the middle part a desiccated level (d) generating to the top breccia fragments (b). The uppermost part corresponds to a rust colored level with shell fragments and bone debris. Sarrebourg, White Beds Formation.

According to Folk & Land (1975), dolomitization is dependent upon Mg++/Ca++ ratio. In hypersaline brines, a high ratio (above 5) is necessary before dolomite forms, because the strong ionic load prevents development of the very well ordered dolomite crystalline lattice. On the other hand in more dilute water, like sea or fresh water, this ratio becomes lower (around 1 to 2) because of a much lower ionic load, allowing slower crystallization and favoring theref Linearity parallel to stratification, of millimetric to centimetric cavities evoke dissolution of previous sulfate nodules diagenetically grown within the dolomitic mud (Figure 12). Silicification occurs at different scales: centimetric nodules distorting dolomitic lamination (Figure 13), scatterad areas of xenomorphic and isomorphic quartz crystals (Figure 14) or even automorphic, showing hexagonal sections. Quartz crystals commonly display microscopic inclusions of minute anhydrite laths obviously related of the presence of anhydrite before silicification takes place (Figure 14). Both diagenetic processes result from the same initial phenomenon: development of Ca-sulfate nodules, later subjected to dissolution or silicification.

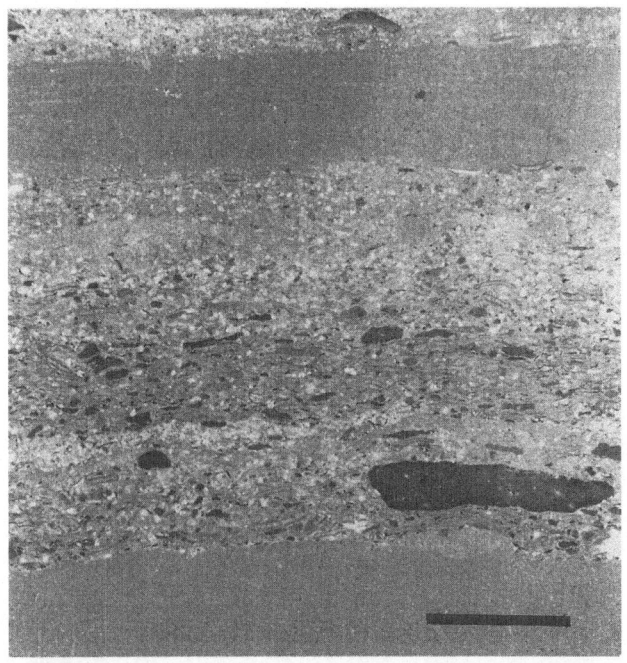

Figure 9. Photomicrograph of a biodetrital packstone layer composed of graded bedded black pebbles, micritized shell fragments and silts interlayered in a dolomicrite. They represent a primary depositional sequence. Sarrebourg, White Beds Formation (plane light, bar = 1 cm).

Some traces of pedogenesis were observed in one instance where, in thin section, a dolomicrite shows very thin, rust colored tubes, looking very much like roots. This episode of probable pedogenesis is contemporary with sedimentation since it is directly topped with a dolomicrite containing shell fragments.

Calcitization affects these dolomites, developing inside silicified areas and as irregular millimetric patches. In some argillaceous facies, calcite extends horizontally along stratification planes and vertically along cracks, generating cellular limestones. These are not observed on cores, only in outcrops; therefore they seem to be connected with very late diagenesis.

Calcium - sulfate facies

They are essentially composed of anhydrite, more or less gypsified. Some gypsum is the result of late diagenesis and occurs as crack filling veins or cemented within argillite.

Gypsified anhydrite facies

Mainly observed in the Grey Beds Formation, anhydrite comes in the form of metric beds showing thin millimetric laminations, regular or contorted (Figure 15).

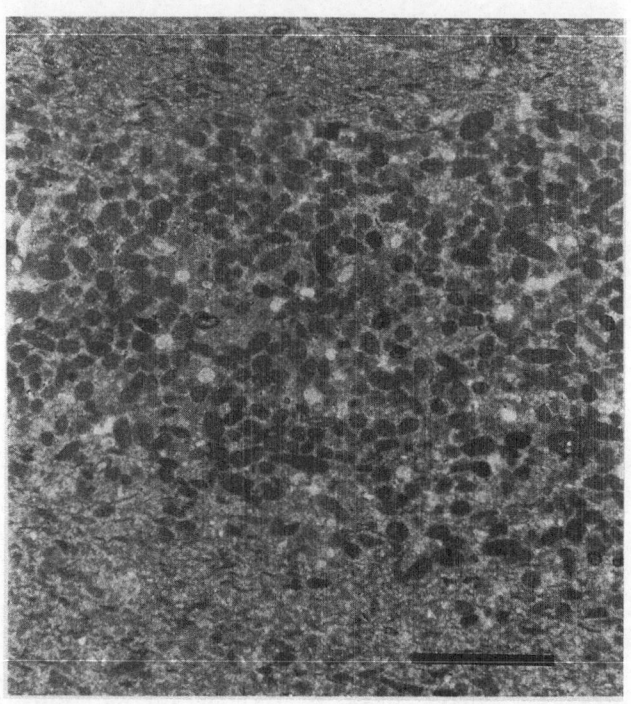

Figure 10. Photomicrograph of a packstone layer composed of pellets associated with small biodetrital fragments and some Miliolideae. Sarrebourg, White Beds Formation (plane light, bar = 3 mm).

Lamination is less obvious in thin sections. However it is possible to distinguish, on an anhydrite or gypsum background, alternating silt and carbonate layers (Figure 16) or even carbonate and sulfate layers (Figure 17), probably representing primary depositional sequences. In some instances millimetric lamination is only emphasized by carbonates. In other cases lamination disappears and a nodular mosaic structure develops (Figure 18).

These various aspects of lamination call to mind algal mats in solar salt works (Geisler-Cussey, 1986) where more or less silty carbonate or gypsum layers alternate with cyanobacterial layers. Organic matter may disappear later and then the entire deposit becomes gypsified, probably, as shown in solar salt works, during a very early diagenesis preserving lamination or destroing it as it passes into a nodular mosaic structure. Later the whole was transformed into

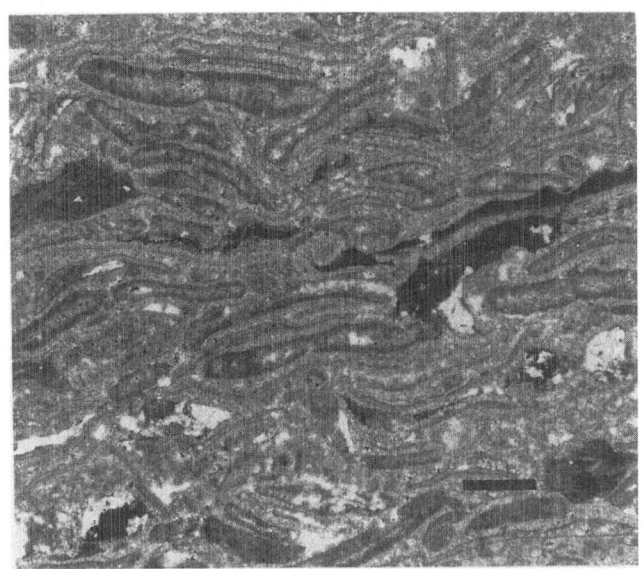

Figure 11. Photomicrograph of a packstone layer composed of micritized shell fragments in a dolomite matrix. Sarrebourg, White Beds Formation (plane light, bar = 2 mm).

Figure 12. Light beige dolomite showing a level of rounded dissolution cavities, resulting probably from dissolution of sulfate nodules. Sarrebourg, White Beds Formation.

anhydrite after burial.

Microscopic observation allows two types of anhydrite crystals to be distinguished, either thick and short (some tens of microns in length) constituting a pure mass, or elongated automorphic (some hundreds of microns in length) generally tightly packed in argillaceous-carbonate matrix (Figure 19). It is worth noting the lathlike shape of anhydrite crystals developing inside a matrix, just like halite which develops a cubic shape in argillites.

All sulfate deposits considered in the present paper come from depths less than 200 m. Therefore they are more or less gypsified because of rehydration processes. Gypsification occurs as scattered gypsum patches showing centimetric poikilotopic crystals. Commonly the deposit undergoes total gypsification.

Late diagenesis gypsum

Large gypsum crystals, scattered or joining together in a pink colored mosaic occur sporadically. Commonly they form veins of satin spar gypsum parallel or sometimes oblique to stratification. This kind of gypsum is undoubtedly generated by late diagenesis in connection with water movements close to the surface. It is due to volume increase of CaSO to CaSO * 2H O and the excess volume is commonly seen as satin spar veins in the same or nearby rocks (Shearman et al., 1972; Mossop and Shearman, 1973).

Halite facies

The halite, studied in the Faulquemont area, occurs in the Grey Beds Formation as bedded salt or displacive halite crystals scattered in an argillaceous matrix. The Red Beds Formation displays only pseudomorphs after halite cubes.

Figure 13. Silicified sulfate nodule in a thin laminated dolomite matrix which is disturbed by early diagenetic nodular growth. Sarrebourg, White Beds Formation.

Figure 14. Photomicrograph in a silicified area of isomorphic quartz crystals showing inclusions of thin anhydrite laths. Sarrebourg, White Beds Formation (plane light, bar = 1 mm).

Figure 15. Grey anhydrite showing a more or less disturbed thin lamination, reminding algal mat structures. Sarrebourg, Grey Beds Formation (bar = 2 cm).

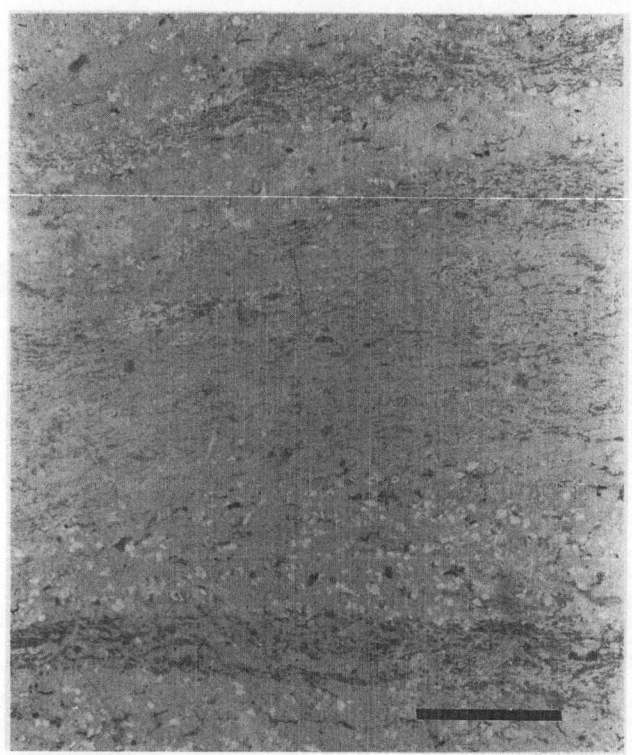

Figure 16. Photomicrograph of alternating silt (white quartz grains) and carbonate (dark) layers in a gypsum matrix, representing primary depositional sequences. Sarrebourg, Grey Beds Formation (plane light, bar = 5 mm).

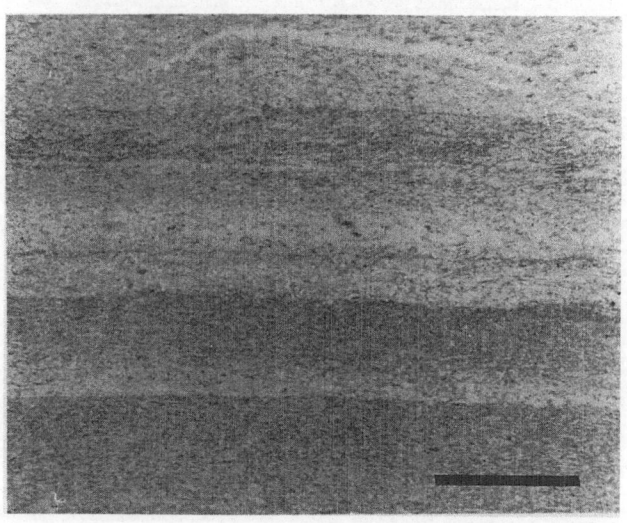

Figure 17. Photomicrograph of alternating carbonate and gypsum layers, representing primary depositional sequences. Sarrebourg, Grey Beds Formation (plane light, bar = 8 mm).

Figure 18. Photomicrograph of nodular mosaic anhydrite showing some dark carbonate relics. Sarrebourg, Grey Beds Formation (plane light, bar = 1 mm).

Figure 19. Photomicrograph of submillimetric anhydrite laths in a dark argillaceous carbonate matrix. Sarrebourg, Grey Beds Formation (plane light, bar = 1 mm).

108

Bedded salt

It occurs as altenating centimetric beds of white and grey salt (Figure 20). White salt is mainly composed of cloudy, inclusion-rich halite and grey salt displays clear halite with many argillaceous impurities, containing millimetric sulfate nodules.

White salt is a primary deposit while grey salt is generated by diagenetic displacive growth of clear halite crystals inside argillite layers. Alternation of both salt types makes up primary depositional sequence,characterized at the bottom by a dilution episode allowing clay deposition, followed by new concentration to brines precipitating halite. The whole undergoes early sulfate and halite diagenesis, consisting in displacive crystal growth from connate brines inside argillites.

Figure 20. Bedded salt showing to the bottom a layer a grey salt with many clay inclusions (gs) and to the top a layer of white salt (ws). The whole represents a primary depositional sequence. Faulquemont, Grey Beds Formation.

Displacive halite in argillaceous matrix

This facies is prevailing in Faulquemont area where many clear halite cubes (about 1 cm an edges) develop in a grey, more or less sulfate rich, argillaceous matrix, showing no bedding (Figure 21). As for grey salt, halite cubes seem to grow displacively from connate brines during an early stage of diagenesis; this hypothesis is supported by the lack of bedding. It is well known that this kind of facies prevailed in the outer rim of Silurian Michigan Basin (Kunasz, 1970) and developed in an anhydrite and dolomite matrix, showing poor bedding (Dellwig, 1955). Holocene continental-sabkha deposits in Bristol Dry Lake (California) also display chaotic mudstone halite on the rims of the playa basin while salt pan halite beds accumulate in its center (Handford, 1982). The same setting occurs in the Middle Muschelkalk of Eastern Paris Basin, because the Faulquemont area is located along the northern edge of the salt deposits.

Pseudomorphs after halite cubes are probably generated in the same way, as scattered displacive halite cubes which underwent later dissolution (Plaziat & Desprairies, 1969). This facies corresponds to a lateral evolution of the previous one in a more dilute environment. Halite pseudomorphs are also described close to the Ardennian shore and therefore characterize the edges of the basin.

FACIES DISTRIBUTION

Except for the detrital argillaceous mass of the Red Beds Formation, Middle Muschelkalk deposits of Lorraine are relatively thin bedded, which is the result of interactions between different kinds of components. Clay mineral supplies, sometimes associated with coarser detrital particles (silt, pellets, shell fragments ...), chemical precipitation and algal growth generate this bedding. These depositional mechanisms have been clearly demonstrated in present solar salt works in southern France (Geisler-Cussey, 1986) and in Spain (Orti Cabo et al., 1984). Therefore the primary depositional seqences are composed of two or three alternating facies. The kind of chemical or organic components present depends on the salinity fluctuations in the depositional environment.

In the Grey Beds Formation, it is difficult to follow facies distribution over the whole area because the significant sulfate diagenesis affecting carbonate deposits masks most of the primary sedimentary structures. However some of them may be sometimes deciphered as previously described.

On the other hand the different facies, essentially argillaceous and dolomitic components differenciate very well on outcrop in the White Beds Formation. Detailed study of the upper 12 m of the series on an outcrop in Sarrebourg area, allows us to distinguish five megaseqences characterized by the succession of green argillites, white argillaceous dolomites and white dolomites (from the bottom up) (Figure 22).

The megasequence itself is divided into decimetric sequences described on a detailed lithologic study of the lowest megasequence at

Figure 21. Displacive clear halite cubes diagenetically grown inside an argillaceous dark matrix containing Ca-sulfate nodules. Faulquemont, Grey Beds Formation.

Sarrebourg outcrop (Figure 22). The first 50 cm are composed of about ten sequences in which green argillites are topped with white argillaceous dolomites; they may also display an intermediate member with alternating millimetric laminae of green argillites and white argillaceous dolomites. The middle portion of the megasequence is characterized by the apperance of a third dolomitic member which continues to the top of the sequence, which is repeated six times in 60 cm. The upper part of the megasequence is mainly dolomitic and starts with an apparent desiccation breccia followed by centimetric alternations of coarse dolomite, which is most evident at the bottom and with a fine dolomite, becoming dominant toward the top. Some rust colored layers, a few millimeters thick, are observed in the coarser dolomite, and show shell fragments, Miliolideae and some bone debris in thin sections. Sometimes these layers appear with an eroded surface at the bottom, indicating some depositional strength.

One of the main features of the megasequential evolution in the White Beds Formation corresponds to the progressive enrichment in

111

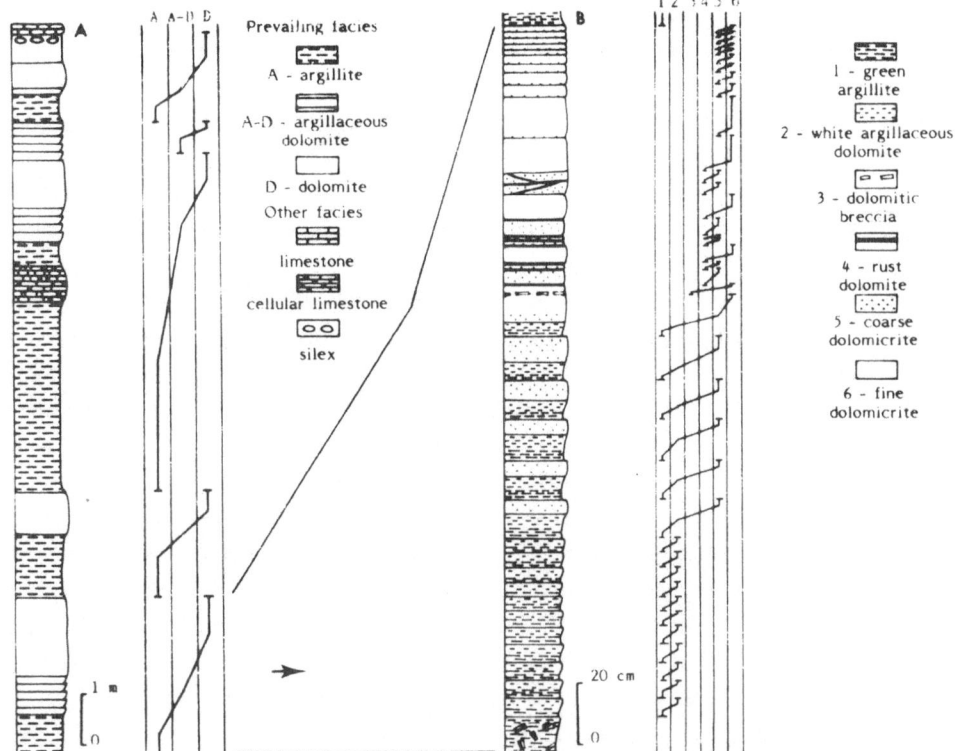

Figure 22. Lithologic section in the White Beds Formation in the Sarrebourg area. A - Lithologic log and megasequential analysis of the entire section, B - Lithologic log and sequential analysis of the lowest megasequence of the section (from Geisler, 1978).

dolomite content of the argillites. Millimetric alternations of green argillites and white argillaceous dolomites, previously described at the bottom of the megasequence, represent primary depositional sequences. They form the transition in decimetric sequences between the green argillite bottom member and the white argillaceous dolomite top member.

GEOCHEMICAL DATA

CLAY MINERAL ASSOCIATIONS AND GEOCHEMISTRY

Clay mineral associations were established from X-ray diffraction along the Middle Muschelkalk section taken from the Sarrebourg bore hole. The same samples underwent chemical analysis by X-ray fluorescence.

Clay   mineral   associations

Four clay mineral associations, ordered from major to minor components, are distinguished: illite-chlorite, illite-chlorite-irregular mixed-layer clay (chlorite-smectite), irregular mixed-layer clay (chlorite-smectite)-illite-chlorite and corrensite-illite. The clays show a progressive evolution from illite-chlorite to corrensite-illite associations through the intermediary step of irregular mixed-layer clays.

Distribution of clay mineral associations vertically through the section is given in Figure 5. The Red Beds Formation is exclusively characterized by the illite-chlorite association. The Grey Beds Formation shows the same clay association at the bottom, but with some irregular mixed-layer clays added further to them. In the White Beds Formation the clay mineral association consists of illite, irregular mixed-layer clays and corrensite. Chlorite is often present, but may disappear and corrensite becomes the main mineral, especially when the sample has a high dolomite content. In the millimetric alternations of green argillites and white argillaceous dolomite, the clay mineral associations are illite-chlorite-irregular mixed-layer clays in the green laminae and corrensite-illite in the white laminae.

Clay   mineral   geochemistry

Major components are analyzed on the silico-aluminium phase, separated from the bulk sample with help of ionic resins (Montanari, Geisler & Petit, 1979). The analytical results plotted on an Al_2O_3 - K_2O - MgO triangular diagram show a linear evolution from an aluminium-potassium area to a magnesium pole, indicating magnesium substitution for the potassium (Figure 23).

Cross plots of clay mineral associations clearly show a magnesium enrichment in connection with appearance of irregular mixed-layer clays. The highest magnesium content is present when corrensite forms. Moreover the section of the Sarrebourg drillings (Figure 5) indicates a stratigraphic distribution of clay mineral change to the top because of magnesium entering in their crystalline lattice.

The same phenomenon is obvious in the White Beds Formation where the millimetric alternations of green argillites and white argillaceous dolomite also show a magnesium enrichment (Figure 23).

Such features are difficult to explain just by primary clay segregation. It seems more likely that the changes result from early diagenesis involving magnesium-enriched connate brines, thrown out of the clay layers because of compaction into more porous carbonate layers. This process allows in the same way magnesium enrichment of clay and carbonate dolomitization. Such a diagenesis may develop layer by layer, preserving primary bedding.

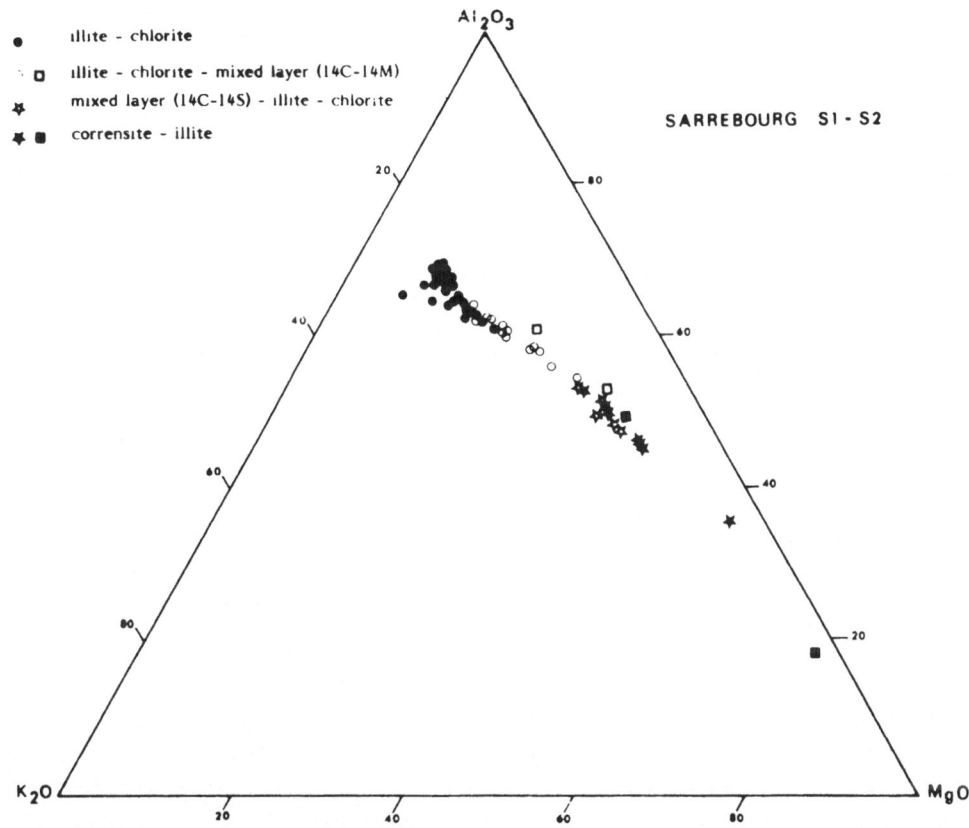

Figure 23. Distribution on a triangular Al_2O_3 - K_2O - MgO diagram of clay mineral associations in argillites of Sarrebourg bore holes. Square symbols correspond to samples taken from millimetric alternations of green argillites and white argillaceous dolomites in the White Beds Formation (from Geisler, 1982b).

BROMINE IN HALITE

The analysed samples come from a bore hole in the Faulquemont area, close to the northern edge of salt deposits, and from some cuttings of Puttelange-aux-Lacs site, located in a more central part of the basin. Unfortunately no other salt samples were available. Sampling was carried out based upon segregation of the type of salt: white or grey; the latter including displacive halite in an argillaceous matrix.

In the Faulquemont samples the bromine contents are lower in white than in grey salt, indicating a lower bromine content in cloudy, inclusion-rich halite of the white salt than in the clear, displacive halite of the grey salt (Figure 24). This result is of great interest because it follows from it that surficial brines generating primary white salt are of lower salinity than connate brines from which clear halite grows displacively. There is obviously a downwards increasing salinity gradient in connection with higher densities.

The bromine contents in halite are in the range of 30 to 70 ppm (Figure 24). In fact it is actually well known from experimental data (Braitsch & Herrmann, 1963) that halite of first precipitation from sea water evaporation has a bromine content of about 70 ppm. Lower values indicate that halite is precipitating from brines having dissolved previously salt, which caused NaCl increase and lower bromine content. Therefore, because its bromine content is lower than 70 ppm, the Middle Muschelkalk salt on the northern edge of the basin was precipitating from mixed mother brines. Salt may have been dissolved by fresh water coming from the Ardennian shore, as shown by the presence of spores, but probably mainly from brines not already saturated with respect to halite. A totally non-marine origin (Hardie, 1984) is difficult to assume, because the bromine contents are not that low (Holser et al., 1972) and the general paleogeography at Middle Muschelkalk time suggests connections with the main German Basin (Ziegler, 1982). Furthermore, the presence of pseudomorphs after halite cubes are a good support for the ongoing dissolution processes. It is also worth noting that bromine contents decrease toward the top of the salt formation, indicating progressive desalination (Figure 24).

In the Puttelange-aux-Lacs bore hole, the bromine content in the halite is in the same range as in the Faulquemont site (Figure 25). This bore hole is located in a more central part of the salt area and allows thought that salt dissolution and represciptation took place not only on the edges of the basin, but as a general phenomenon in the Middle Muschelkalk of eastern Paris Basin.

DYNAMIC BEHAVIOR OF THE BASIN

Middle Muschelkalk of Lorraine in eastern Paris Basin is just a small evaporitic sub-basin of the southwestern edge of the large German Basin. The distribution of evaporites is controlled by the structural behavior of the Hercynian NE-SW substructure. For example the halite is restricted to the area of main subsidence, which corresponds to the present Sarreguemines syncline, while the deposits become thinner in areas of lesser subsidence, such as in the present Sarro-Lorrain anticline. It is also possible to note a slight shift in subsidence toward the west, which is the first step toward the individualization of the Paris Basin. This pattern of the basement subsidence is evidenced mainly during the deposition of evaporites of the Grey Beds Formation and seems to have been less active before and after, during sedimentation of the Red Beds and White Beds Formations.

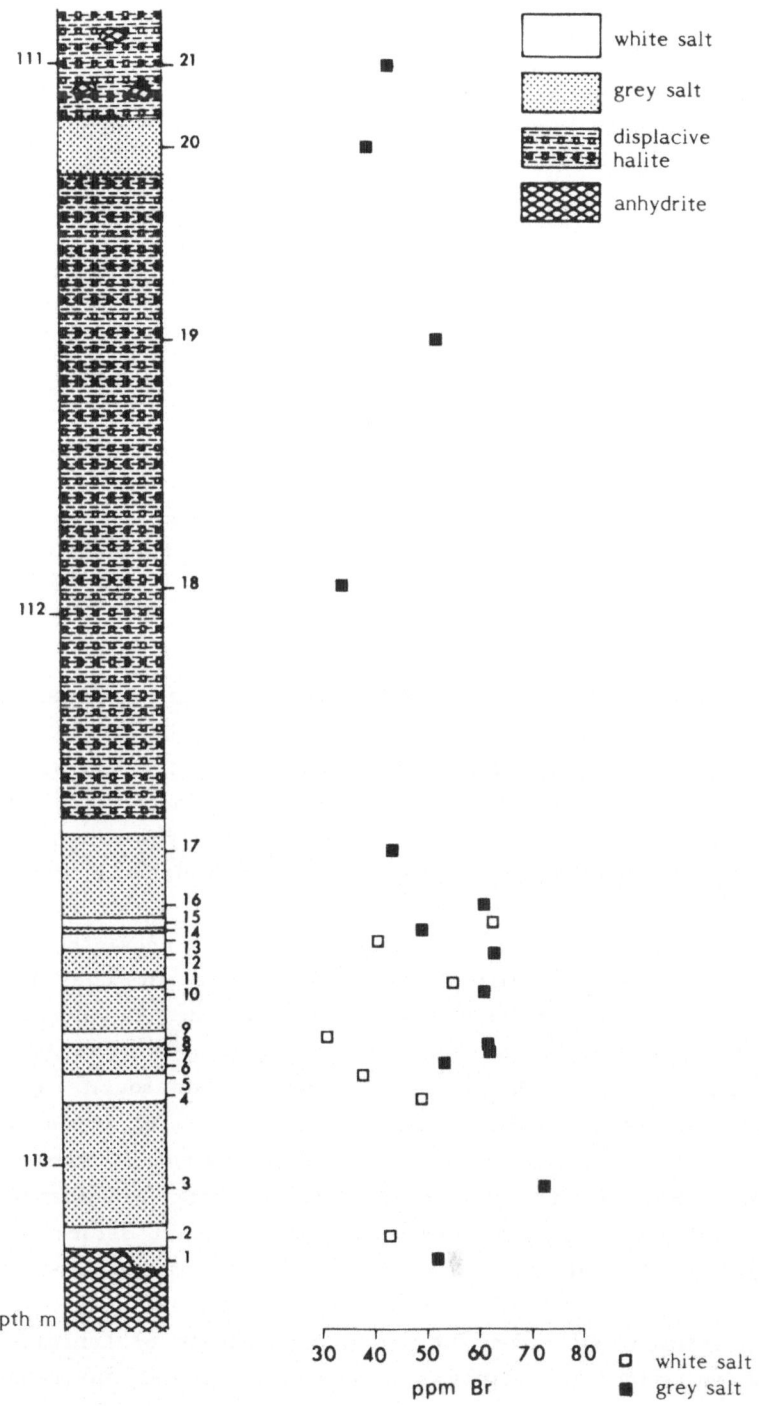

Figure 24. Lithology of the salt level in the Faulquemont bore hole and bromine contents in halite (from Geisler, 1982b).

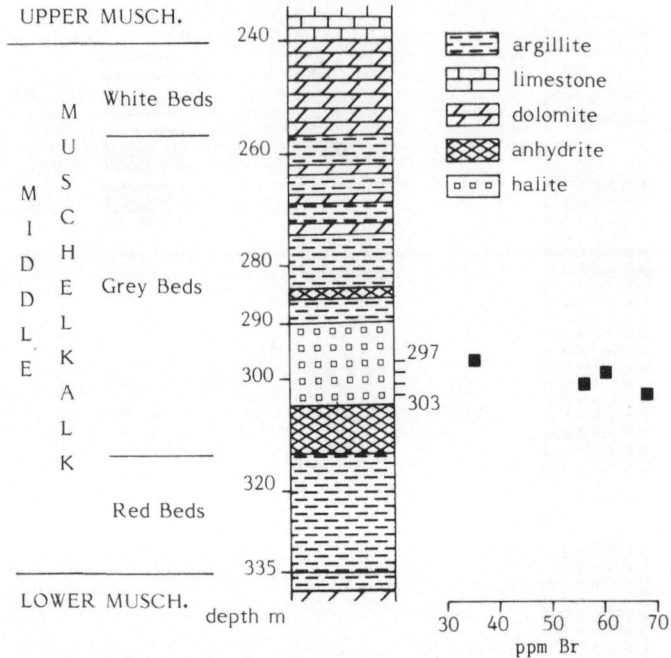

Figure 25. Log of Puttelange-aux-Lacs bore hole and bromine contents in halite cuttings from the salt formation (from Geisler-Cussey, 1986).

Palynological data permit us not only to give an Upper Anisian age to the series, but also to establish that there were probably no significant climatic changes, because the palynological association maintains with the same composition throughout the entire Middle Muschelkalk series, in this area.

The closeness of the Ardennian continent provided fresh water supplies, as shown by the presence of spores in the palynological associations on the northern edge of the basin. Furthermore, it fed the basin with detrital material, relatively coarse close to the shore and at the bottom of the series further to the south. However, the continental influence becomes less and less important moving away, from the Ardennian massif since there is no equivalent of the argillaceous Red Beds Formation in the Middle Muschelkalk of Wurttemberg in Germany and also a lesser amount of clay in that series (Wild, 1973). In Lorraine the detrital supply is fine grained and relatively homogeneous, mainly composed of illite and chlorite clay particles.

The sedimentary record of Middle Muschelkalk of Lorraine indicates a cyclic evolution of salinity in the depositional environment. For instance, the Red Beds Formation is characterized by fine detrital argillaceous material settling in waters which were more and more concentrated, as shown by gypsum occurrences towards the top of the formation. Grey Beds Formation contains two Ca-sulfate levels flanking a middle salt level which correponds to the highest salinity. Finally, the White Beds Formation marks a salinity decrease with the apperance of dolomite, before return to normal marine conditions when the Upper Muschelkalk transgression takes place.

From correlations with Triassic of the German Basin (Gall, Durand & Muller, 1977), the cyclic evolution of salinity could be connected from the beginning of Middle Muschelkalk, with Tethyan water supplies through the Silesian sill (Kozur, 1975); that means a connection to a seaway which is at a great distance. This allows considerable brine concentration before reaching eastern Paris Basin. On the other hand, at the end of this period the general transgression in the German domain, resulting probably from a significant eustatic change, caused progressive dilution in the Lorraine area.

The middle of the salinity cycle (its maximum) is located in the Grey Beds Formation and corresponds to the salt deposit. The bedded salt is restricted to the area of main subsidence at the center of the basin which is probably deeper, allowing brine stratification and therefore salt preservation. Towards the edges of the basin, the salt deposit is only generated from diagenetic growth of halite cubes within an argillaceous mud. The cubes show an increasing sparcity and finaly dissolve near the edges, forming pseudomorphs after halite. This evolution indicates exposure during periods of lower water towards the edges of the basin, preventing halite preservation.

Bromine contents in halite are lower than is usual at the begining of halite precipitation from normal evaporated sea water, but not neglegible, indicating salt recycling through fresh water from the Ardennian continent, but mostly from influx of unsaturated sea water. These results tend to demonstrate that the Middle Muschelkalk salt was rarely protected by saturated brines and therefore was affected by dissolution-reprecipitation procesess. Therefore the basin was probably not very deep and the brines were not commonly stratified.

Ca-sulfate deposits of the Grey Beds Formation generated before and after salt precipitation are not actually primary sediments, but rather products of very early sulfate diagenesis of thin laminated carbonates similar to those formed by algal mats. They suggest a massive cyanobacterial settlement in the basin. The algal growth is periodically interruped by fine detrital influxes, by carbonate or sometimes by gypsum precipitation. These processes have been observed in present solar salt works on the Mediterranean coast of southern France (Geisler, 1982a). The water depth is probably low, but the deposits display no evidence of aerial exposure. Further the whole sediment undergoes gypsification with/without preservation of the sedimentary structures, just as it has been described in present solar salt works of Spain (Orti-Cabo et al., 1984). After burial, gypsum is transformed into anhydrite. The very good lateral correlation of such sulfate beds, generated by diagenesis, remains something of a problem.

The depositional environment of the White Beds Formation shows a clear evolution of the water to lower salinites. The presence of shells of molluscs, ostracods and Miliolideae interlayered in dolomites, supports periodic inflow of normal sea water into the basin. The water in which the White Beds formed, shows shallowing with some episodes of aerial exposure, as shown by desiccation breccias and pedogenetic processes.

Magnesium enrichment in the interstitial brines causes an important early diagenesis in the White Beds Formation. This enrichment is mainly restricted to the carbonate layers at the top of the depositional sequences, where dolomitization develops and magnesium rich clays, such as corrensite, are generated. Such an early diagenesis preserves the sedimentary structures. The magnesian-rich character of sedimentation disappears suddenly in the Upper Muschelkalk with deposition of normal marine limestones.

The present late diagenesis in the Middle Muschelkalk of eastern Paris Basin is related to the entry of meteoric waters. It develops zones of dissolution and fissure filling with satinspar gypsum.

CONCLUSIONS

Middle Muschelkalk of the eastern Paris Basin is an extension of the large German Basin than an unit by itself. It allows us to study the effects of continental influences on evaporitic sedimentation: fresh water and detrital supplies, and salt dissolution. More or less evaporated sea water flowing into the basin causes a cyclic evolution of salinity throughout the series.

The deposit is generally layered and shows millimetric to centimetric primary depositional sequences which prove the complexity of the internal organization of the filling in the basin. They reflect elementary changes in the depositional environment: fresh water supplies carring detrital particles, fresh water and sea water inflow introducing dilution, increasing evaporation rate causing precipitation of carbonates, Ca-sulfate or halite. In fact, the main factors controlling sedimentation which are working at this fine scale and the evolution at a larger sequential scale depends upon which one of these factors becomes dominant. The general evolution results in a typical sedimentary cycle.

The structural behavior of the Hercynian basement controls distribution and sometimes preservation of the deposits. Diagenesis may also preserve or totally destory primary sedimentary structures. Algal mat-like deposits undergo extensive Ca-sulfate diagenesis in the middle of the sedimentary cycle and toward its top, dolomites and magnesium-rich corrensites develop layer by layer.

Middle Muschelkalk of the eastern Paris Basin is a very good example for the study of sedimentary processes involoved on the edges of a large evaporitic basin.

ACKNOWLEDGEMENTS

The author would like to thank Drs. T.Hilly and G.Busson for many helpful suggestions and Dr. B.C.Schreiber for critical reviewing the manuscript.

REFERENCES

Adloff M.-C., Doubinger J. & Geisler D., 1982 - Etude palynologique et sedimentologique dans le Muschelkalk moyen de Lorraine. Aspects stratigraphiques, paleoeceologiques et paleogeographiques. Sci. de la Terre, XXV, no. 2, 91-104.

Braitsch O. & Herrmann A.G., 1963 - Zur Geochemie des Broms in salinaren Sedimenten. Teil I: Experimentelle Bestimmung der Br Verteilung in verschieden naturlichen Salzsystemen. Geochimica et Cosmochimica Acta, 27, 361-391.

Dellwig L.F., 1955 - Origin of the salina salt of Michigan. J. Sedim. Petrol., 25, 83-110.

Folk R.L. & Land L.S., 1975 - Mg/Ca ratio and salinity; two controls over cristallization of dolomite. AAPG Bull., 59, 60 - 68.

Fourmentraux J., Pontalier Y., Lavigne J. & Poujol P., 1959 - Trias, Jurassique inferieur et moyen de l'Est du Basin de Paris. Presentation des cartes d'isopaques et de lithofacies. Rev. I.F.P. et Ann. Combustibles Liquides, XIV, 1063-1090.

Gall J.-C., Durand M. & Muller E., 1977 - Le Trias de part et d'autre du Rhin. Correlations entre les marges et le centre du bassin germanique. Bull. B.R.G.M., serie 2, section IV, no. 3, 193-204.

Geisler D., 1978 - Une coupe detaillee dans le sommet du Muschelkalk moyen a Sarrebourg (Moselle). 103. Congr. nat. Soc. sav. Nancy, fasc. IV, 335-341.

Geisler D., 1982a - De la mer au sel: les facies superficiels des marias salants de Salin-de-Giraud (Sud de la France). Geol. Medit., IX, 521-549.

Geisler D., 1982b - Muschelkalk moyen de Lorraine. Donnees geometriques, sedimentologiques et geochimiques. Sci. de la Terre, XXV, no. 2, 71-90.

Geisler-Cussey D., 1986 - Approche sedimentologique et geochimique des mecanismes generateures de formations evaporitiques actuelles et fossiles. Marais salants de Camargue et du Levant espagnol - Messinien mediteraneen et Trias lorrain. Mem. Sci. de la Terre, no. 48, 268 pp.

Handford C.R., 1982 - Sedimentology and evaporite genesis in a Holocene continental-sabhka playa basin - Bristol Dry Lake, California. Sedimentology, 29, 239-253.

Hardie L.A., 1984 - Evaporites: marine or non-marine? Am. J. Sci., 284, 193-240.

Holser W.T., Wardlaw N.C. & Watson D.W., 1972 - Bromide in salt rocks: extraordinarily low content in the Lower Elk Point salt Canada; In: G. Richter-Bernburg (ed.), Geology of Saline Deposits: Paris UNESCO, 73-76.

Kozur H., 1975 - Probleme der Triasgliederung und Parallelisierung der germanischen tethyalen Trias. Teil II: Anschluss der germanischen Trias an die internationale Triasgliederung. Freiberger Forschungshefte, Leipizg, C 304, 51-77.

Kunasz I.A., 1970 - Significance of lamination in the Upper Silurian evaporite deposits of the Michigan Basin. Third Symposium on Salt, Northern Ohio Geol. Soc., 1, 67-77.

Laugier R., 1959 - Observations petrographiques nouvelles sur les niveaux saliferes du Trias moyen de Lorraine. Bull. Soc. Geol. Fr., serie 7, I, 31-39.

Le Roux J., 1971 - Structures tectoniques et anomalies gravimetriques dans l'Est de la France. Bull. B.R.G.M., serie 2, section 1, no. 3, 137-141.

Maget P. & Maiaux C., 1980 - Trias. Facies evaporitiques. Mem. B.R.G.M., no. 102.

Marchal C., 1983 - Le gite salifere keuperien de Lorraine-Champagne et les formations associees. Etude geometrique. Implications genetiques. Sci. de la Terre, XXIII, no. 44, 139 pp.

Montanari R., Geisler D. & Petit G., 1979 - Mise au point d'une methodologie analytique concernant certains types de roches evaporitiques. Sci. de la Terre, XXIII, no. 1, 3-27.

Mossop G.D. & Shearman D.J., 1973 - Origins of secondary gypsum rocks. Trans. Inst. Mining and Metallurgy, B 82, 147-154.

Orti Cabo F., Pueyo Mur J.J., Geisler-Cussey D. & Dulau N., 1984 - Evaporitic sedimentation in the coastal salinas of Santa Pola (Alicante, Spain). Rev. Ins. Invest. Geol., 38/39, 169-220.

Plaziat J.-C. & Despraires A., 1969 - Les pseudomorphoses de cristaux de sel gemme du Keuper inferieur de Lorraine: mode de formation et repartition paleogeographique. Bull Soc. Geol. Fr., serie 7, XI, 400-406.

Richter-Bernburg G., 1972 - Saline deposits in Germany: a review and general introduction to the excursions. In: G. Richter-Bernburg (ed.), Geology of Saline Deposits, Paris, UNESCO, 275-287.

Ricour J., 1962 - Contribution a une revision du Trias francais.Mem. Carte Geol. France, 471 pp.

Shearman D.J., Mossop G.D., Dunsmore H. & Marin M., 1972 - Origin of gypsum by hydraulic fracture. Trans. Inst. Mining and Metallurgy, B 81, 149-155.

Trusheim F., 1971 - Zur Bildung der Salzlager im Rotliegenden und Mesozoikum Mitteleuropas. Beih. Geol. Jb., 112, 1-51.

Wervecke van L., 1916 - Die Kustenausbildung der Trias am Sudrande der Ardennen. Mitt. Geol. Landesanstalt Els.-Lothr., X, 151-239.

Wild H., 1968 - Das Steinsalzlager des Mittleren Muschelkalks, seine Entstehung, Lagerung und Ausbildung nach alter und neuer Auffassung. Jb. Geol. Landesamt Baden Wurttemberg, 10, 133-155.

Wild H., 1973 - Neue Erkenntnisse uber Genese und Lagerung des Salzes im Mittleren Muschelkalk in Suddeutschland. Ober. und Mitt. oberrh. Geol.Ver., Stuttgart, N.F. 55, 95-132.

Wurster P., 1964 - Geologie des Schilfsandsteins. Mitt. Geol. Staatsinst. Hamburg, Heft 33, 140 pp.

Ziegler P.A., 1982 - Geological Atlas of Western and Central Europe. Shell International Petroleum Maatschapij, Elsevier, 130 pp.

SEDIMENTARY MODELS OF GYPSUM-BEARING CLASTIC ROCKS AND PROSPECTS FOR ASSOCIATED HYDROCARBONS WEST OF THE TARIM BASIN (CHINA) IN MIOCENE

Qiu   Dongzhou

Comprehensive Institute of Petroleum and Geology
Ministry of Geology and Mineral Resources
Jiangling, Hubei
China

INTRODUCTION

The geometry of the west of the Tarim Basin, approximately 100 000 sq. km in area, was crosswise-bell-like in Miocene. It was bounded to the north by the Tian (Heaven) Mountains and Bachou Hills, to the south by the Kunlun Mountains, and to the east by the western part of Tarim Basin. It was connected with the ancient Dula Sea of Middle Asia intermittently in the west. Wuqia Formation (Miocene) consists of dominantly interbedded brown gypsiferous sandstones and mudstones, intercalated with greyish green mudstones and conglomerates, usually 2,000 - 3,000 m thick (maximally 7,000 m). In broad terms, three formations, in ascending order, can be differentiated: Keziluoyi Formation (N1k), 500 - 800 m thick, composed dominantly of sandstones, intercalated with mudstones and gypsum; Anjuan Formation (N1a), 500-1,000 m thick, consisting of mudstones, intercalated with sandstones and gypsum; and Pakabula Formation (N1p), 1,200 - 2,000 m thick, built by interbedded sandstones and mudstones, with intercalations of conglomerates and gypsum.

Observation in field and analyses in laboratory suggest that principal sedimentary environments in the early and middle Miocene differed from not only normally continental lake basins but also common bays and lagoons, being a kind of continental lake-basin with a few characters of marine environment. This environment once was invaded by the sea for a moment. In contrary, principal sedimentary environments in later Miocene were normal interior lakes.

SEDIMENTARY ENVIRONMENTS AND MODELS IN EARLY AND MIDDLE MIOCENE

MAIN SEDIMENTARY ENVIRONMENTS AND FEATURES

Four facies are recognized in this period: lake delta, fluvial, alluvial fan, and transgression lake. The first three facies are similar to those which are commonly described. Transgressive lake facies possesses the features of common lake facies. Moreover, the signs of transgression can be found. These include:

Petrology-Mineralogy

(1) Gypsum deposits developed widespread. Their contents in section is up to 1 - 2%. Halo-sylvite, carnallite, and glaserite are contained in gypsum, but no alkali is found. Evaporites related to marine facies can be seen.

(2) Foraminifers-bearing biogenetic oolitic limestones developed in some area. The biogenetic oolitic limestone of Anjuan Formation (Yangye, Wuqia), for example, has a total thickness of 35 cm, consists of seven strata, each stratum 2 - 8 cm thick. The components are foraminifers, gastropods, and ostracods. The tests of foraminifers are well preserved, and no transportation traces have been observed.

(3) The diagram of size parameters shows that the most samples belong to the transitional area between beach and fluvial facies. Y values of fourteen of sixteen samples exceed 7.42, they can be interpreted as shallow sea facies, on the basis of Sahu's formula for environments differentiation.

(4) Glauconite can be seen in the rock section from Ke weel-4, Kekeya oil-field. It shows the features of cement and is considered to be primary.

(5) En echelon beddings and tittle beddings consisting of lenticular beddings, wavy beddings, and flaser beddings were developed in Anjuan Formation at Keziluoyi Ravine, Waqia.

Paleontology

A number of *Ammonia beccaris* and *Cyprideis littoralis* was recorded. *Ammonia beccaris*, throught studies in the Yellow Sea, the Eastern China Sea, and Mexico Gulf, usually lives in a brackish transitional environment including littoral facies, bay, lagoon, and estuary. The depth of water is about 10 - 20 m. The other features are:

(1) Marine and continental organisms occur together. The upper member of Anjuan Formation at Keziluoyi Ravine, for instance, contains not only *Ammonia beccaris* and *Cyprideis littoralis* living in

littoral area but also eucypris, condona, and charophyta.

(2) The species composition is monotonous, and euryhaline organisms propagated vigorously. Foraminifers found in Anjuan Formation at Keziluoyi section consist almost of *Ammonia beccaris*, 60% of ostracods being *Cyprideis littoralis*. And 90% of ostracods is *Cyprideis littoralis* in Wulukekati profile.

(3) Variations in species occurred widespread. At least six species of variant *Ammonia beccaris* are identified. All mentioned above are considered to be developed in a brackish transitional and transitional lake envinronment.

Geochemistry

(1) Trace element analysis shows that the average content of boron in Anjuan Formation is 92 ppm, and the ratio of barium to gallium is 3.4 - 4.2. These values are interpreted to be characteristic of transitional envinronment.

(2) The content of potassium and sodium in Anjuan Formation is up to 62 and 182 ppm, respectively, from Maichan Well-1 in Bachou.

(3) The geochemical analysis of source rocks shows a strong predominance of hydrocarbons at C20 - C24, 1.4 - 1.7 for the ratio of Pr and Ph. This indicates that the type of kerogen is transitional between type I and type III. Thus, it is referred that organic matter in Anjuan Formation, Wuqia was deposited in a transitional environment.

DEPOSITIONAL MODELS

A sedimentary model in early and middle Miocene is presented in Figure 1 and Figure 2, referring to material of Tertiary deposits in the Estern China, contemporary Maracaibo and Taihu Lake. The following features can be concluded:

(1) The south flank of Kunlun Mountains uplifted and compressed obviously. Peripheral facies, including piedmont fans and fluvial facies, having a larger thickness, developed. However, tectonization in north flank was rather weak. Structures in middle part were relatively steady, and transitional lake deposits developed.

(2) The deposits in the south are coarser and thickner, and in the north are finer and thinner. The deposition centre and subsidence were inconsistent.

(3) On the basis of the features of deposition, paleontology, and geochemistry, the suggestion is that the depth of semi-deep and shallow transgresion lakes was 10 - 20 m and 0 - 10 m, respectively.

(4) Transgression lakes extended east and west. Thicker deposits spread from east to west. All these features indicate that studied area was once connected with ancient Dula Sea off and on. The sea invaded

from the west by channel had an important influence on deposition features in early and middle Miocene.

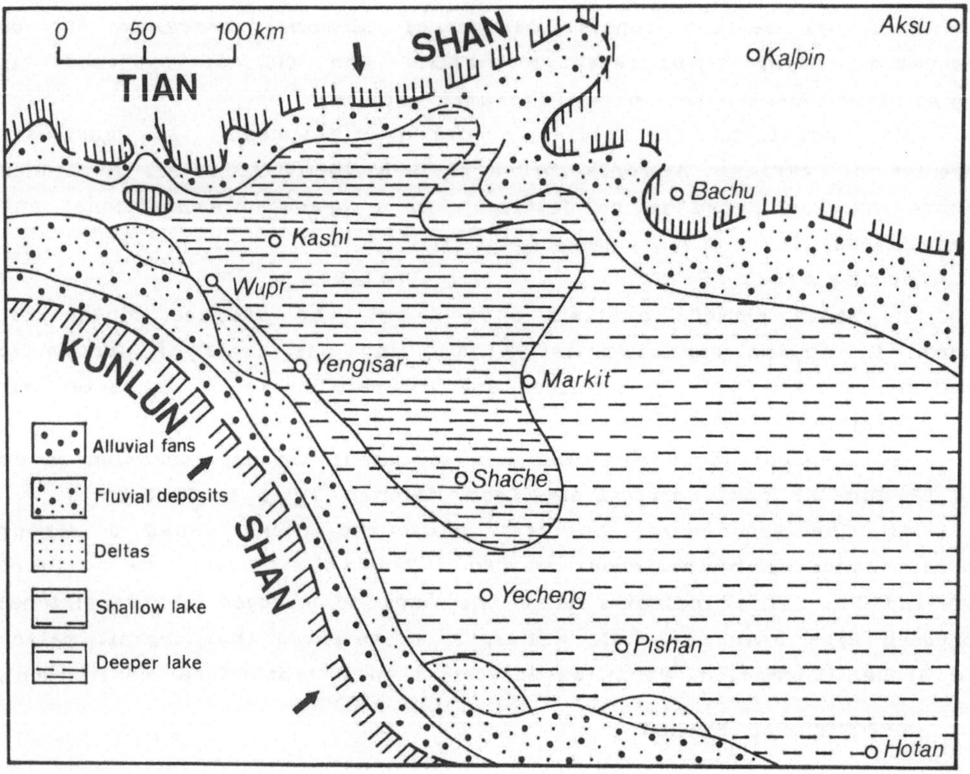

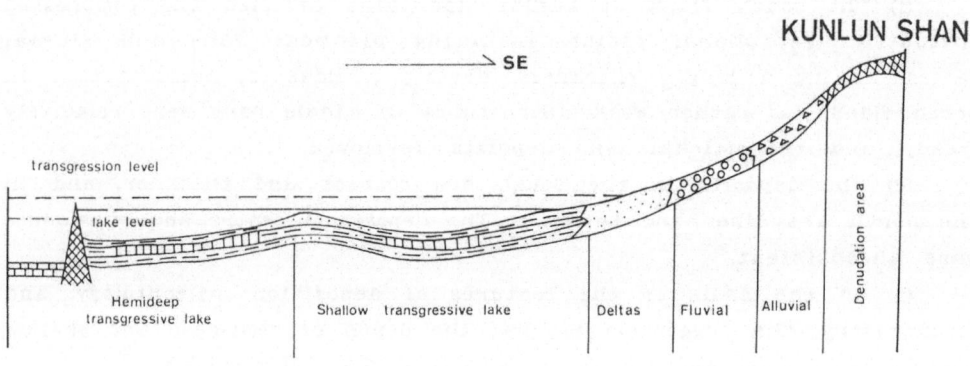

Figure 1. Sedimentary facies sketch west of the Tarim Basin during early and middle Miocene.

Figure 2. Scheme of facies relations west of the Tarim Basin during early and middle Miocene.

(5) Two periods of transgression can be differentiated : the first occurred during the deposition of Keziluoyi Formation and was short and the affected range was small: the second, in middle Miocene was relatively larger and wider, spreading to Qimonggen, Maigaiti, and Qukuqiake.

SEDIMENTARY TYPES AND MODELS IN LATER PERIOD

MAJOR TYPES OF SEDIMENTARY SETTING AND FEATURES

Five environments can be recognized in later Miocene: interior lake, lake delta, fluvial and lake turbidity current. The last one is most particular, and is characterized below.

(1) Lithological association was intrerbedded sandstones and mudstones. Sandy bodies are relatively steady and can be correlated in the entire oil field.

(2) According to statistics of Ke Well-51, -7, and the core of Pa no. 4 and no. 5 member in Ke Well-9, the content of greenish gray basic igneous rock, white quartzite and vein quartz, purple acid igneous rock, parti-colored sand-mudstone, and carbonate in conglomerate rocks is 30%, 25%, 10% and 10% respectively. And 65 - 75%, 7 - 10%, 14 - 18%, 1 - 3%, and 10 - 5% for quartz, feldspar, debris, mica, matrix, respectively, are characteristic in sandstone section. They are arcose-lithic quartz siltstone and matrix-bearing arcose-lithic quartz siltstons. A complex rock component indicates deposition near terrigenous province and rapid accumulation.

(3) The sorting and roundness of rocks are poor, and composition maturities are low. Different size of grains, from cobble to silt, can be seen in a core 26 cm long, from Ke Well-4. The shapes of conglomerate rocks are subangular to subrounded.

(4) In grain size analysis, probability curves show a low slope angles and cover a wider grain size.

(5) Throught observation of cores from eight wells, three combinations of bedding can be distinguished. The first is combination of A-B-C-D bedding; the second is of A-B-C bedding; and A-B bedding for the third (Figure 3). Among these A,B,C,D means graded beding, evenly bedding, wavy bedding or wavy-cross bedding or deformed bedding, horizontal bedding or uncontinuously horizontal bedding, respectively.

6) Self-potential logs can be differentiated into two parts. The lower parts, bell-like, with the benth of curves changing abruptly, change upward from larger amplitude to smaller amplitude. The upper parts, funnel-like, changing to the curves of superstrata smoothly, change upward in a contrary tendency. In comparison to the curves of

Tertiary turbidite in Huanghua depression, many common features can be found among them.

(7) Lacustrine ostracods such as *Cyprinotus*, *Cypris*, *Eucypris*, and fish fossils are contained in horizontally bedded mudstones. They indicate the water body was rather deep.

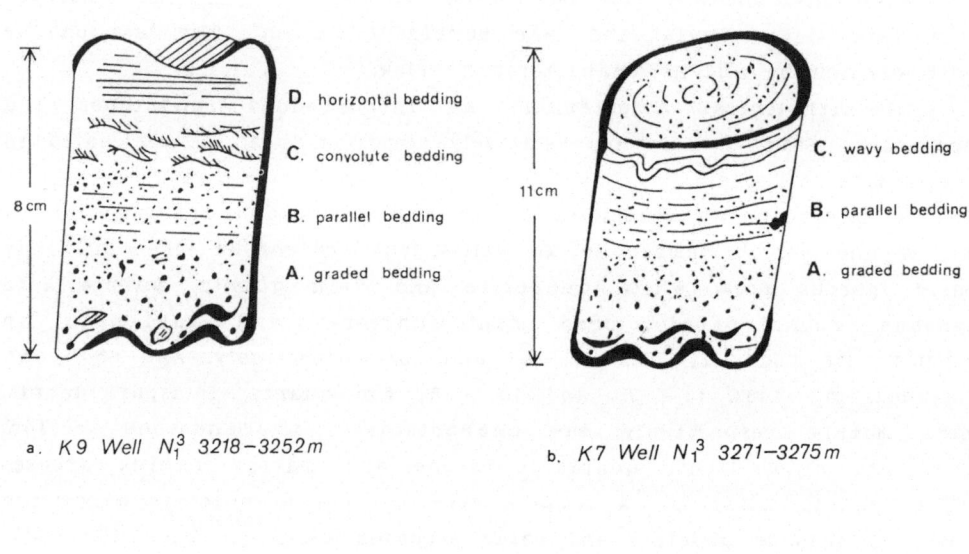

a. *K9 Well N$_1^3$ 3218-3252m*

b. *K7 Well N$_1^3$ 3271-3275m*

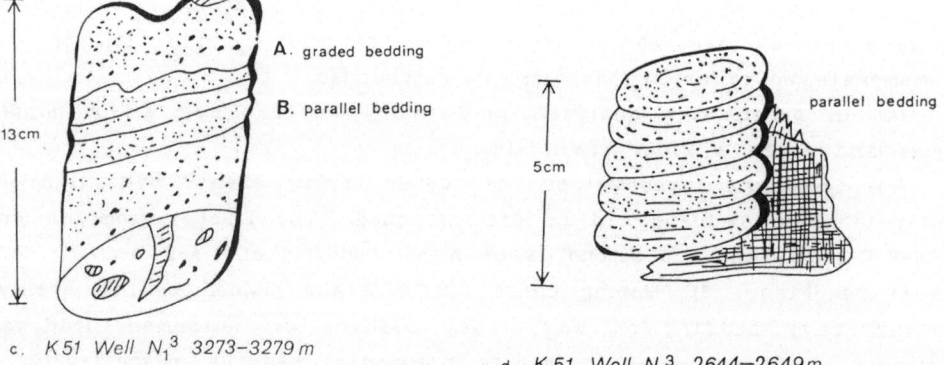

c. *K51 Well N$_1^3$ 3273-3279 m*

d. *K51 Well N$_1^3$ 2644-2649m*

Figure 3. The combination of beddings found in cores of eight wells.

THE GENERATING CONDITIONS OF TURBIDITE OF PA NO. 4 AND NO. 5
MEMBER IN KEKIYA OIL-FIELD

(1) The matrix content of turbidite-bearing lake deposits is 5 -
10%, 15 - 20% for the corresponding fluvial and delta deposits in Pusa.
And matrix content in fluvial and delta deposits of Momoke, south-west
of Pusa, exceeds 20%. An increase in density of deposits in an internal
factor which induces turbidity to develop (Figure 4).

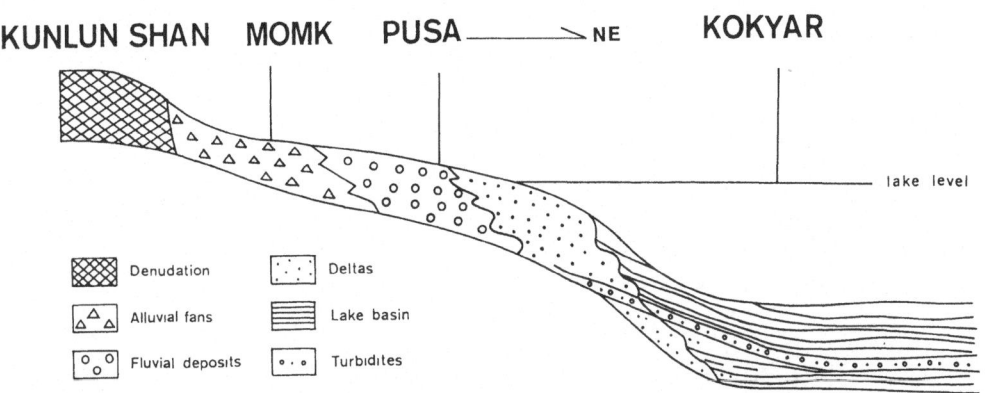

Figure 4. Scheme of facies relations in Pa no. 4 and no. 5 member
within Kekeya oil-field.

(2) The thickness of Kabulake Formation in Momoke, Pusa, and
Kekeya oil-fields is 1,603m, 1,651m, and 2,322 m, respectively. On the
ground of seismic cross-sections, the formation in Kekeya is thicker
than that in Pusa area. This implies that there was an initial slope
from the south to the north while Kabulake Formation formed. This
initial slope favoured the generation of turbidity flows from south to
north.

(3) Taking a number of earthquakes and floods occurring along the
piedmont of contemporary Kunlun Mountains into consideration, it is
suggested that two factors have a contribution to the development of
turbidity currents in Pakabulake stage of Miocene.

Having studied the generation conditions and characteristics of Pa
no. 1 and no. 5 member turbidity deposits in Kekaya oil-field, the
conclusion is that they belong to terrigenous clast lake turbidity
current. At the same time, they have the features of being near
terrigenous province and developing some fan-deltas.

It should be noticed that the uplift of Kunlun Mountains in later
Miocene was evident. The steep flank of dustpan-like basin served as a

good geological setting for fan-delta turbidity flow generation. In accordance with field observations and grain-size analyses, the suggestion is there exists similar turbidides in Pakabulake Formation of Yiliya Ravine, Hesilapu, and Duwa.

SEDIMENTARY MODEL

The sedimentary model for later Miocene are reconstucted in Figure 5 and Figure 6. From those figures is evident that:

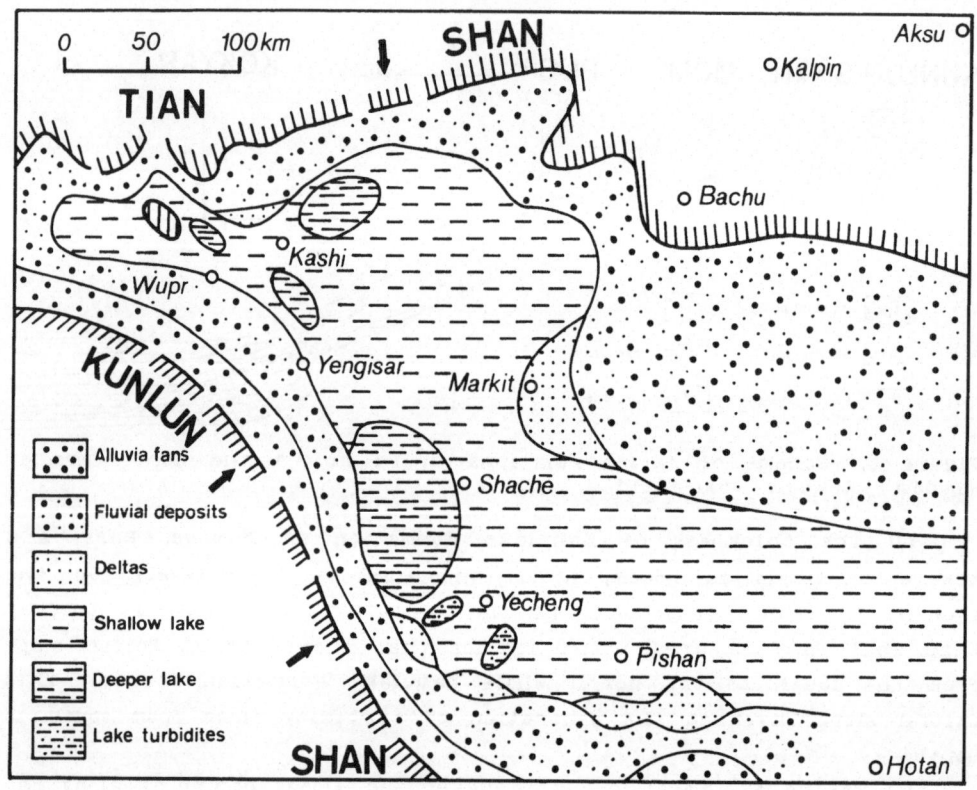

Figure 5. Sedimentary facies sketch of the area located west of the Tarim Basin in later Miocene.

(1) The features of tectonic movements in terrigenous province and depressed area in latter Miocene are similar to those in early and middle Miocene. The tectonic movements in Kunlun Mountains are more evident than in Heaven (Tian) Mountains. The marginal facies developed well, and thicker deposits spread along the piedmont of Kunlun Mountains while poorly developed and thinner deposits are characteristic for the piedmont of Tian Mountains. The nature and patterns of tectonic movements in terrigenous province still controlled obviously sedimentary facies and thickness.

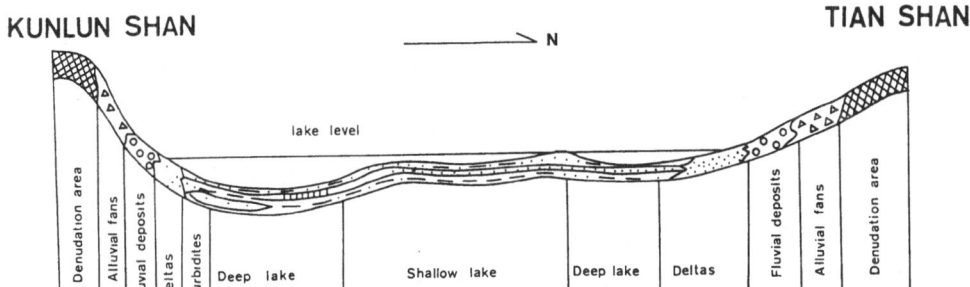

KUNLUN SHAN   ⟶ N   **TIAN SHAN**

Figure 6. Scheme of facies relations west of the Tarim Basin in later Miocene.

(2) The distribution of semi-deep lake interior was different from that of semi-deep transgression lakes in early and middle Miocene. It was not on the middle part of lakes but on the fringe of lakes. For example, these differences can be found in Yiliyasi Ravine, Keketamu, and Pakabulake.

(3) Considering sedimentological, paleontological, and geochemical data, the suggestion is that the depth of lakes interior and shallow lakes interior was about 10 m and 0 - 10 m, respectively. They had the characteristics of fluvial lakes.

(4) With the uplift and compression movements in Paermei area, Kunlun Mountains were raised. The channel which connected the west part of Tarim Basin with ancient Dula Sea was sealed. Thus, the history of connecting with the ancient Mediterranean was ended.

(5) Because of the obvious uplift movements of Kunlun Mountains, there was an apparent difference between the height of piedmont and that of lake basins. A large amount of terrestrial deposits formed lake turbidite with some features of fan delta when their density exceeded that of overlying water.

SEDIMENTARY MODELS AND ASSOCIATED HYDROCARBONS

Based of the sedimentary models of gypsum-bearing clastic rocks west of the Tarim Basin, following suggestions on oil and gas exploration in the west of the Tarim Basin results:

(1) Semi-deep transgression lakes are advantageous to the generation of oil and gas. The organic geochemical analyses prove that semi-deep transgression lakes are favourable to the preservation of

organic matter and changing into hydrocarbons.

(2) The shoal and sand bodies in transgression lakes, and interior lake deposits, braid channels and river mouth bars in delta deposits, channel bars and point bars in fluvial deposits spread extensively. Sandstones laid down in these facies, with a larger thickness, relatively pure composition, moderate to fair sorting and roundness, are fair reservoirs.

(3) The pellite and gypsum deposited in transgression lakes and interior lakes, with a larger thickness, can be considered to be good seal. Otherwise, an especially large-scale over-thrust nappes, Kezikeaer, along Kunlun Mountains, probably controlled, during the entire Miocene, the development of pelitic gypsum and gypsum.

(4) The constructional deltas formed while water body of lake basins reduced and fluvial action increased. Prodelta facies, delta-front facies, and delta-plain facies are recognized in stratigraphic sections from bottom the top. It had a large-scale volume and stretched to the interior of basins. It, thus, is a favourable combination for oil and gas accumulation.

(5) The lake turbidite, with large thickness and relative steadiness, moderate physical properties , distributed from basin margin to the interior of basin. Because of being near source rocks, surrounded by lake deposits, many traps formed easily. This suggests that sand bodies in lake turbidites have a fair hydrocarbon-producing potential.

(6) The lake basin appeared dustpan-like in early and middle Miocene. The deposition centre was identical with the depressed centre in plan. But the overlying area, being near source rocks, located at slope in structure, are regionally favourable areas for oil and gas accumulation. As examples, Dongsai, Heizi, Yongjisa, and Wupaer which are located at the piedmont of Kunlun Mountains, may be referred to.

(7) Considering the relationship with hydrocarbons, it seems that the gypsum deposits in Miocene can be divided into two types. The first type developed in semi-deep transgression lake environments and it has much to do with oil-producing rocks, such as found in Anjuan Formation at Kangxiweier and Kashi. The second type is associated with deposits of shallow interior lakes and as was found in Anjuan Formation at Pusa, has little to do with source rocks.

This paper is a summary of "Study on the area west of the Tarim Basin in Miocene" by Comprehensive Institute of Petroleum and Geology. Many thanks are due to my colleagues who have given me much help either in field or laboratory work.

REEF-STROMATOLITES-EVAPORITES FACIES RELATIONSHIPS FROM MIDDLE MIOCENE EXAMPLES OF THE GULF OF SUEZ AND THE RED SEA.

MONTY C.L.V.*, ROUCHY J.M.**, MAURIN A.***,

BERNET-ROLLANDE M.C.***, PERTHUISOT J.P.****

INTRODUCTION.

Most of evaporite basins are characterized by thick evaporitic sections in their central parts (gypsum - anhydrite, halite, K and Mg salts) replaced shoreward by massive reefal carbonates associated with stromatolites. Classical examples have been described from the Messinian of the mediterranean Basin (ESTEBAN, 1979; BERNET-ROLLANDE *et*

  * C.A.P.S., Labo. de Biosédimentologie, Université de Liège, B-4000 Liège, Belgium.
 ** UA 1209 (CNRS) and GREDOPAR, Labo. de Géologie Museum d'Histoire naturelle,75005 Paris, France.
*** TOTAL, C.F.P., La Defence, 75739 Paris Cedex 15, France.
**** GREDOPAR, Dept. des Sces de la terre, Université de Nantes, 44072 Nantes Cedex, France

Lecture Notes in Earth Sciences, Vol. 13
T.M. Peryt (Ed.), Evaporite Basins
© Springer-Verlag Berlin Heidelberg 1987

134

al, 1980; ROUCHY, 1982; ROUCHY *et al*, 1986), the Eocene of the pyrenean Basin (ORTI-CABO *et al*, 1984),the Mississippian of the Artic (DAVIES, 1977),, the Silurian of the Michigan (MESOSELLA *et al*, 1974; HUH *et al*, 1977 ...) etc. In many cases however, the time and genetic relationships between bioconstructed carbonates and evaporite deposits

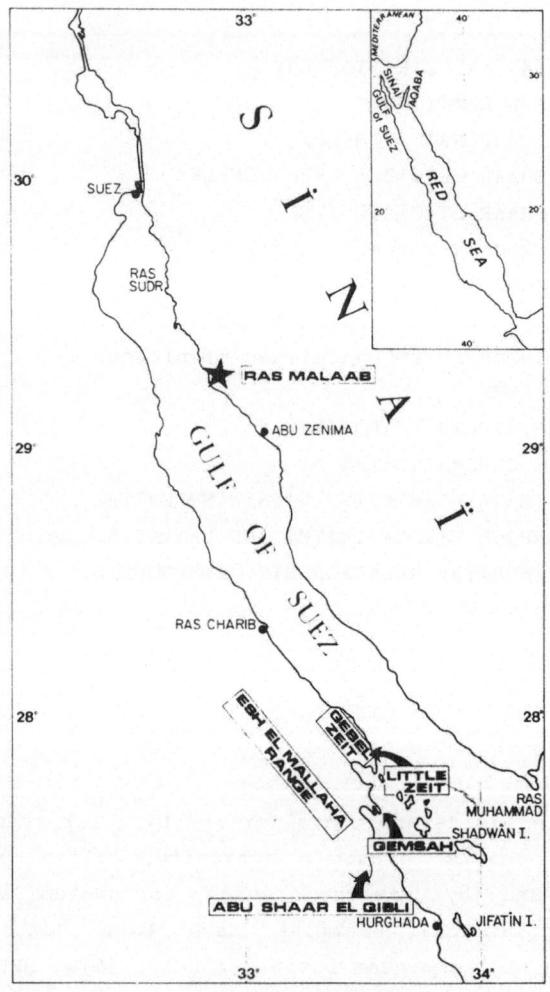

Figure 1. Location of main localities reported in this study.

Plate 1. Satellite view of discussed area; on each side, rough geological maps of Gebel Zeit (upper right) and Gebel Esh Mellaha - Abu Shaar el Qibli complexes (Lower left corner).

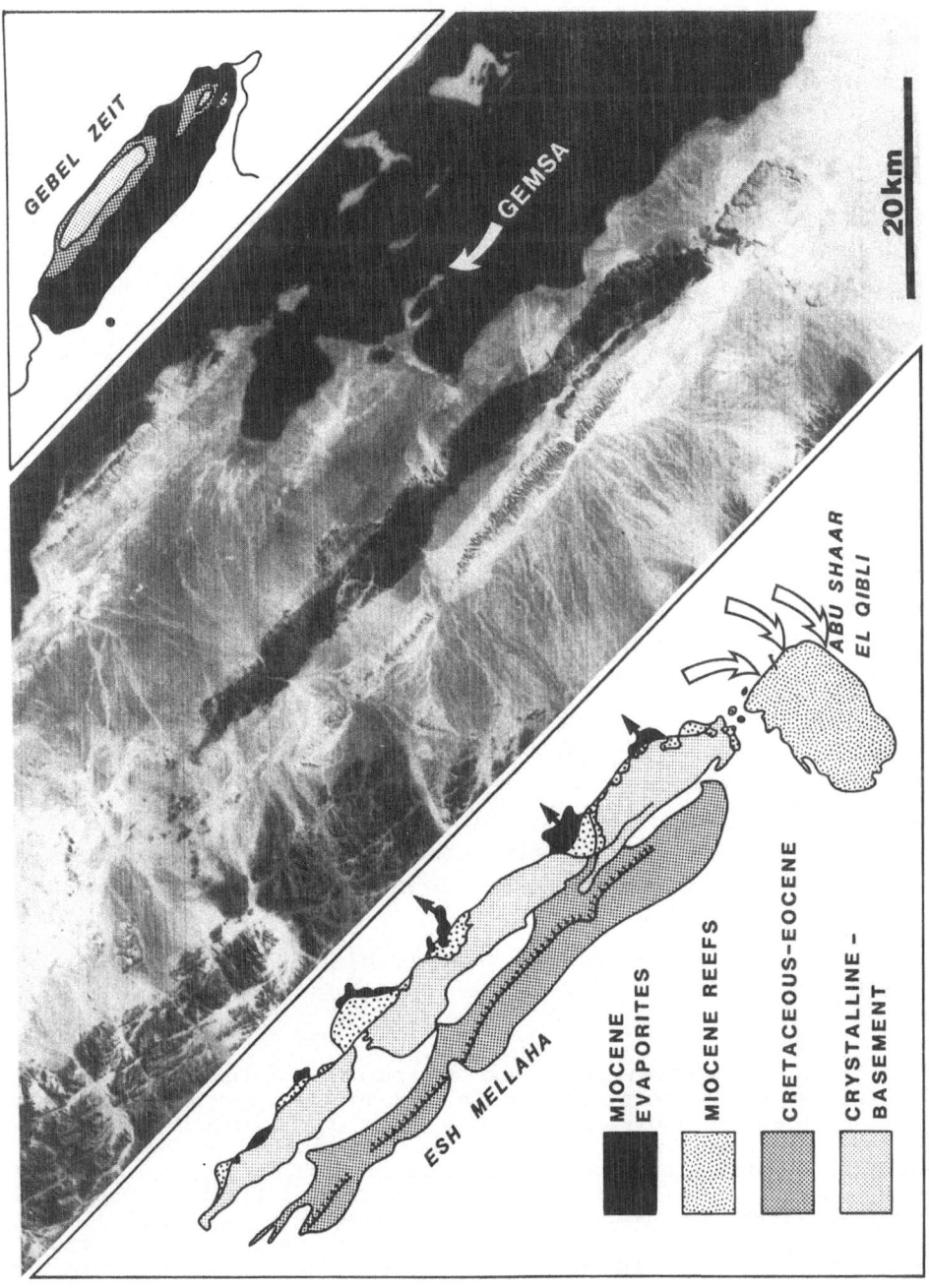

GEBEL ZEIT

GEMSA

20km

ABU SHAAR EL QIBLI

ESH MELLAHA

MIOCENE EVAPORITES

MIOCENE REEFS

CRETACEOUS-EOCENE

CRYSTALLINE - BASEMENT

have been the subject of conflicting interpretations, a situation which reflects deep controversies about the genesis of the thick evaporitic deposits (KIRKLAND and EVANS, 1973).

Analysis of the reef complex fringing the Middle Miocene evaporitic depressions of the Gulf of Suez-Red Sea rift system allows a reconstruction of geometric and genetic relationships between the three major kinds of sediments : reefs, stromatolites and evaporites.

Middle Miocene evaporitic formation up to 3600 meters in thickness (HASSAN and EL DASHLOUTY, 1970) infill individual grabens of the Gulf of Suez - Red Sea rift. Well developed carbonate complexes cap tectonic highs around evaporitic basins, specially along the Esh EL Mellaha Range and Gebel Zeit (Figure 1; Plate 1). The Abu Shaar el Qibli carbonate complex, surrouding the southern end of Esh el Mellaha (Figure 1; Plate 1), has been originally described by MADGWICK et al (1920). It basically consists of a basal coral reef core interfingering backward with bioherms and well bedded lagoonal deposits whereas it passes seaward to and/or interfingers with steep talus facies; this complex is capped by a composite stromatolitic reef blanket (ROUCHY, 1979,1982; ROUCHY et al, 1983, EL HADDAD et al, 1983/1984). Such stromatolitic facies also invades parts of talus slopes. Basinward, lie thick evaporites (1000 m. of gypsum, anhydrite, halite) deposited in a narrow contiguous trough. Stromalitic accretions can be observed within evaporites as well as within marly beds intercalated between gypsum ones, on both sides of the rift (Sinai and Western Desert).

The observed relationships between reef, stromatolites and evaporites appear to bear interesting general implications susceptible to provide an adequate model at least useful for approaching other situations characterized by poor outcropping conditions.

Plate 2. A.- Laminated marls and diatomites intercalated within two gypsum beds; arrows point toward the base of the upper gypsum bed. Note black carbonate/diatomite laminites near the contact with gypsum; southern part of Gebel Zeit. Hammer for scale. B.- Close up of laminites showing thin intercalations of crystalline gypsum. Hammer for scale. C.- Thin section (negative) of calcitized laminated and locally nodular sulfates resulting from microbial sulfate reduction of gypsum and/or anhydrite in presence of organic matter. Note globular porosities replacing vanished micronodules (black patches), Gebel Zeit - Scale bar 2mm. D.- Close up of thin section (negative) in irregular, roughly laminated calcitized sulfate in relation with dense microbial mats (will be explicitated in stromatolite section;see also Plate 14). Main porosity results from presence of elongated ellipsoidal cavities after vanished sulfates. Gebel Zeit.Scale bar 500μm.

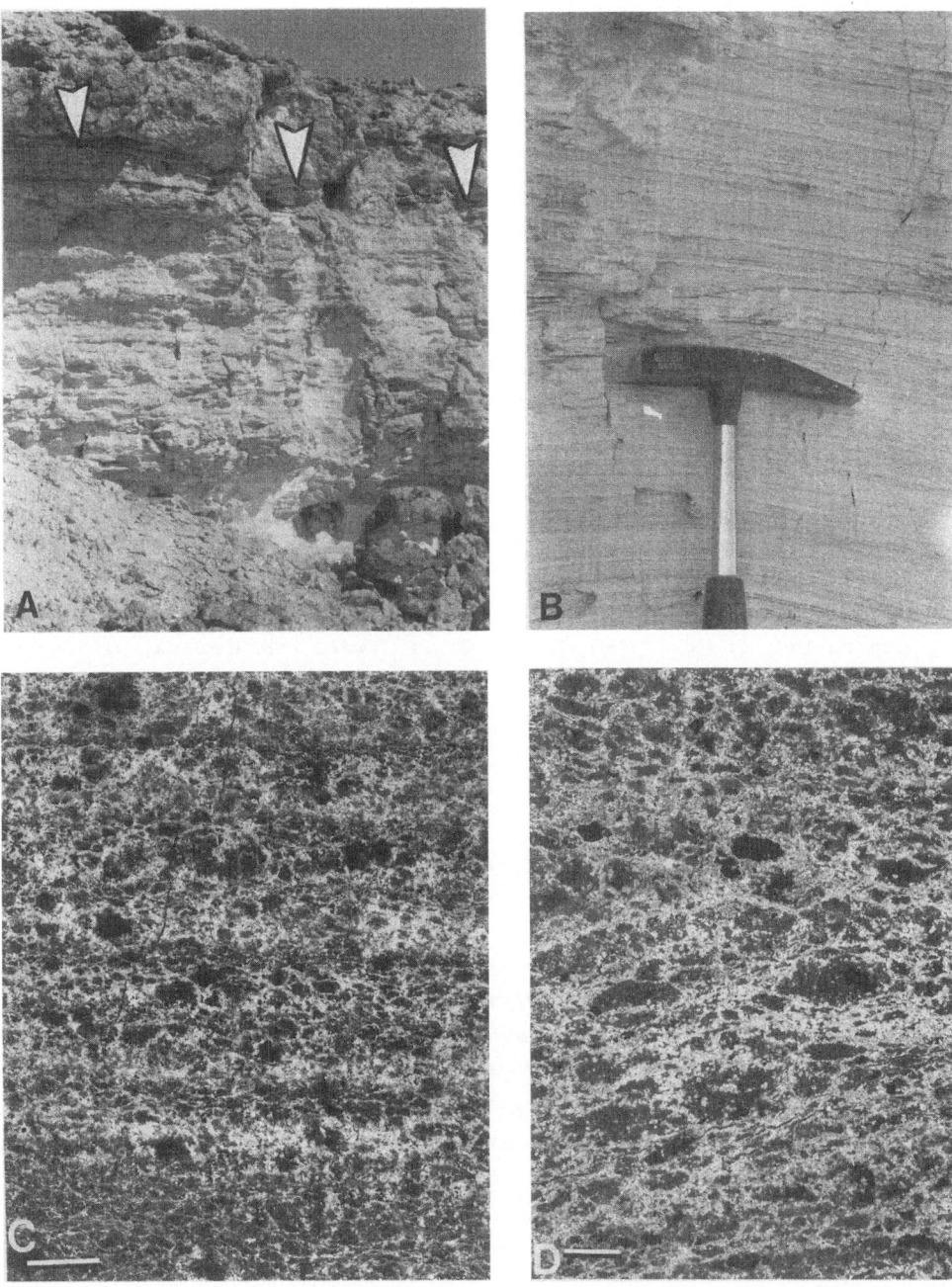

I. GENERAL SETTING AND STRATIGRAPHY.

The regional tectonic framework is characterized by a dominant set of NW-SE. normal faults system which delimits parallel horsts and grabens; this structural pattern was set up during Oligocene - Lower Miocene times in association with the opening of the Gulf of Suez - Red Sea rift system.

Indications of older similar directional distensive tectonic have been locally observed, although slightly oblique to the present rift. The evaporitic formations contributed to infilling of elongated troughs (grabens). The complete filling up sequence begins with continental detrital beds (Nukhul Formation), overlain by thick globigerinid-rich marls of the Rudeis Formation (Burdigalian). In the central part of the graben, the evaporites were laid down into four successive formations (Kareem, Belayin, South Gharib and Zeit) begining during Langhian times (KERDANY, 1968; CRAVATTE and DUFAURE, 1979, written com.; EL HEINY and MARTINI, 1981) and possibly extending upward into the Messinian Stage as controlled in the Red Sea D.S.D.P. well (BOUDREAUX, 1973). A possible extension into Pliocene or Plio-Quaternary (Zeit Bay wells and outcrops) is under investigation. Horsts capped by carbonate complexes parallel the grabens axes.

II. THE EVAPORITIC FORMATIONS.

The series is extremely variable in thickness and composition; from subsurface data, the first anhydritic beds of the Kareem Formation overlie the marine Rudeis marls (Burdigalian-Langhian) equivalent to the *Globigerina*-marls outcropping at Gebel Zeit which are of Langhian age in their upper part (CRAVATTE and DUFAURE, 1979, written com.). Deposition of thick halitic bodies, located in the Belayim and South Gharib Formations, probably started as early as Serravalian

Plate 3. A.- Typical and contrasting sulfate deposit: thinly laminated at the base of photo, nodular in the upper half. Gemsa peninsula. Scale bar is 10cm. **B.-** Relics of vertically stacked selenite crystals converted into anhydrite through climatic alteration. Gemsa Peninsula. Scale bar is 10cm.

times. The Zeit Formation represents a comprehensive series of alternating gypsum, marls and arenitic beds extending into the Pliocene.

Near the edges of uplifted blocks (Gebel Zeit horst, Gemsa diapiric dome), the evaporites are represented by rhythmically alternating gypsum-anhydrite units, up to several decameters in thickness and marly-diatomites (Plate 2,A) The subsurface threefold-division of the lower part of the evaporitic sequence (Kareem, Belayim, South Gharib) can barely be recognized on outcrops where halite may be replaced by massive sulfates; the upper part of Zeit Formation, on the contrary, is characterized by thinner beds. At Gebel Zeit, collapse breccias could be equivalent to the subsurface halitic bodies.

Also, sulfate layers (5-30 cm thick) are here interbedded with decametric laminated, marly, diatomitic sediments, sometimes grading into very pure diatomites (Plate 2,A). The general laminated pattern of marls and diatomites is most probably indicative of episodically stratified and poorly oxygenated waters. Diagenetic tridymite cristobalite spherules after diatoms can be observed (Noel and Rouchy in progress). Some layers however are composed of well preserved but poorly diversified assemblages of diatoms (*Thalassionena*, *Coscinodiscus*, accessory *Aptinotychus*) locally associated with abundant *Orbulina* along with a scarce calcareous nannoplancton and marine fishes which have been discovered in the northern (Ras Dîb) and central parts

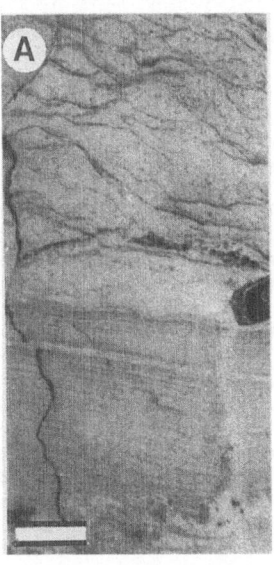

of the Gebel Zeit (Gaudant and Rouchy, in press).

In the upper part of outcropping evaporites, marls, sands and limestones appear to be impoverished, even devoid of marine fauna except for very few mollusc-rich beds (oysters, gastropods, Gebel Zeit). The fossil content hence marks a clear transition from marine conditions for the marls, to restricted, hypersaline conditions. Although observation on halite facies are not sufficient we can suggest that the deposition of the thick halite bodies is mainly subaqueous within residual brine bodies. Futhermore, the presence of potassium salts (GHORAB et al, 1969) attests the installation of episodic phases of hyperconcentration of brines requiring the evaporation of most of water column. The overall observations show that the deposition of the evaporitic formation took place during a period of sea level fluctuations (see below).

Beside diagenetic carbonatation (Plate 2, C, D) that will be discussed later, the sulfates exhibit several facies : coarse crystalline selenite, laminated gypsum/anhydrite, and nodular to mosaic anhydrite (Plates 2, B; 3, A); the first two facies are tentatively interpreted as primary subaqueous (SHEARMAN and ORTI-CABO, 1978 ; SCHREIBER, 1978; ROUCHY, 1982). SELLWOOD and NETHERWOOD (1984) oppose the laminated sulfates that they interpret as of subaqueous origin, to nodular and chicken wire ones which would be indicative of a sabkha origin. If we agree with the presence of evaporitic, supratidal environments on the margins of blocks during lowstands, most of the sulfates appear to be of subaqueous origin. Such sabkha-like deposits would be limited in time and space to ephemeral intertidal to supratidal flats.

Plate 4. A.- Dark diagenetic carbonate bodies postruding at the summit of the outcrop through partly eroded sulfate layers (see close up in E). Note striking similarities with the carbonate buttes described by KIRKLAND and EVANS (1976) in Permian Castille Formation. Similar carbonate bodies although smaller are frequently observed in Gebel Zeit; heigth of the carbonate buttes: 10-15 meters. Sharm el Bahari, 50 km south of Quseir. B.- Close up of diagenetic carbonate showing subrounded centimetric open cavities. Fully cemented cavities do not appear at this scale. Sharm el Bahari. Hammer for scale. C.- Laminoid fenestral fabric partly cemented or eventually floored by internal sediments. Sharm el Bahari. Scale bar: 2cm. D.- Endostromatolite (arrow) encrusting rim of the upper part of decimetric cavity. Sharm el Bahari. Scale bar: 2cm. E.- Close up of the upper part of a diagenetic body such as outcropping on photo A. Note, one the lower half, thinly bedded diagenetically altered marls and laminites underlying a massive vacuolar lens of secondary carbonate, and overlying another one not visible on photograph. Height of outcrop: about 10 m. Sharm el Bahari.

Where cropping out, gypsum is often replaced by anhydrite (Plate
3,C) (climatic dehydration of gypsum), this transformation destroyes
primary structures and produces thin blanket of white, fine-grained
anhydrite (ROUCHY et al, 1986), except in running wadis on recently
activated cliffs or in quarries. Similar examples of surficial
alteration of gypsum under hot, dry climatic condition have been
reported by MAIOLA and GLOVER (1965), SHEARMAN (1971), etc (see also
Plate 3,B).

In some areas (Gebel Zeit, Gemsa, Sharm el Bahari near Quseir,
Wadi Gharandal in Sinai) anomalously-shaped carbonate bodies, decimeter
to decameter thick appear, to result from diagenetic bacterial
reduction of sulfates (ROUCHY et al, 1985 ; PIERRE and ROUCHY, 1986);
they are exposed as discontinuous bodies scattered into massive
sulfates (Plate 4,A,E), at the laminite/sulfate boundary or in the cap
rock of diapirs , a situation reminiscent of that described by KIRKLAND
and EVANS (1976) in the Permian Castille Formation

These biodiagenetic carbonates present a porous to cavernous
structure. Porosity may appear as irregular centimetric subrounded vugs
(Plate 4,B) irregularly scattered in the massive carbonate, as laminoid
fenestrae (Plate 4,C), open or cemented by acicular, clear or blackish,
calcite (sometimes aragonite); in blackish cements, needles form around
central, dark organic filamentous remains ("microbial spars", MONTY,

Plate 5. A.- General view (negative) of elongate diagenetic cavity in
process of being cemented by endostromatolites growing centripetally
from the walls of cavity; note radial microstructure of
endostromatolite (details in C-E); residual voids are separated by
dolomite and some sulfate in association with microbes (see detail in
C-E). Sharm el Bahari (Quseir). Scale : 5 mm. **B.**- Endostromatolites in
outcrop. Scale: 1cm. **C.**- Thin section in outer part of
endostromatolite; one clearly sees here the radiating pattern of
constitutive filaments (central part of photograph) where microcavities
persist between some of them, aggradation of dolomite in the lower part
of the photograph obscured the radial fabric. This growth is capped by
closely spaced micritic dolomitic films. Scale bar 400 μm. **D.**- Close up
of a porous zone in the radial fabric showing vertical growth of long
filaments encrusted by dolomite; note that the filament keep on growing
through the dolomicritic tangential films. Scale bar: 200μm. **E.**- Close
up of D, showing dark microbial filaments (arrow) supporting and
encrusted by a chain of superposed euhedral dolomite crystals. Cross
polarized light shows the common orientation of crystals. Filament grow
towards the right. Scale bar: 100μm. **F.**- Iron impregnated *Tubiphytes*-
like microbial colony in compact replacement dolomicrite. Base of the
colony is along the right hand side of the photograph. Arrow indicates
direction of vertical growth. Scale bar: 200μm.

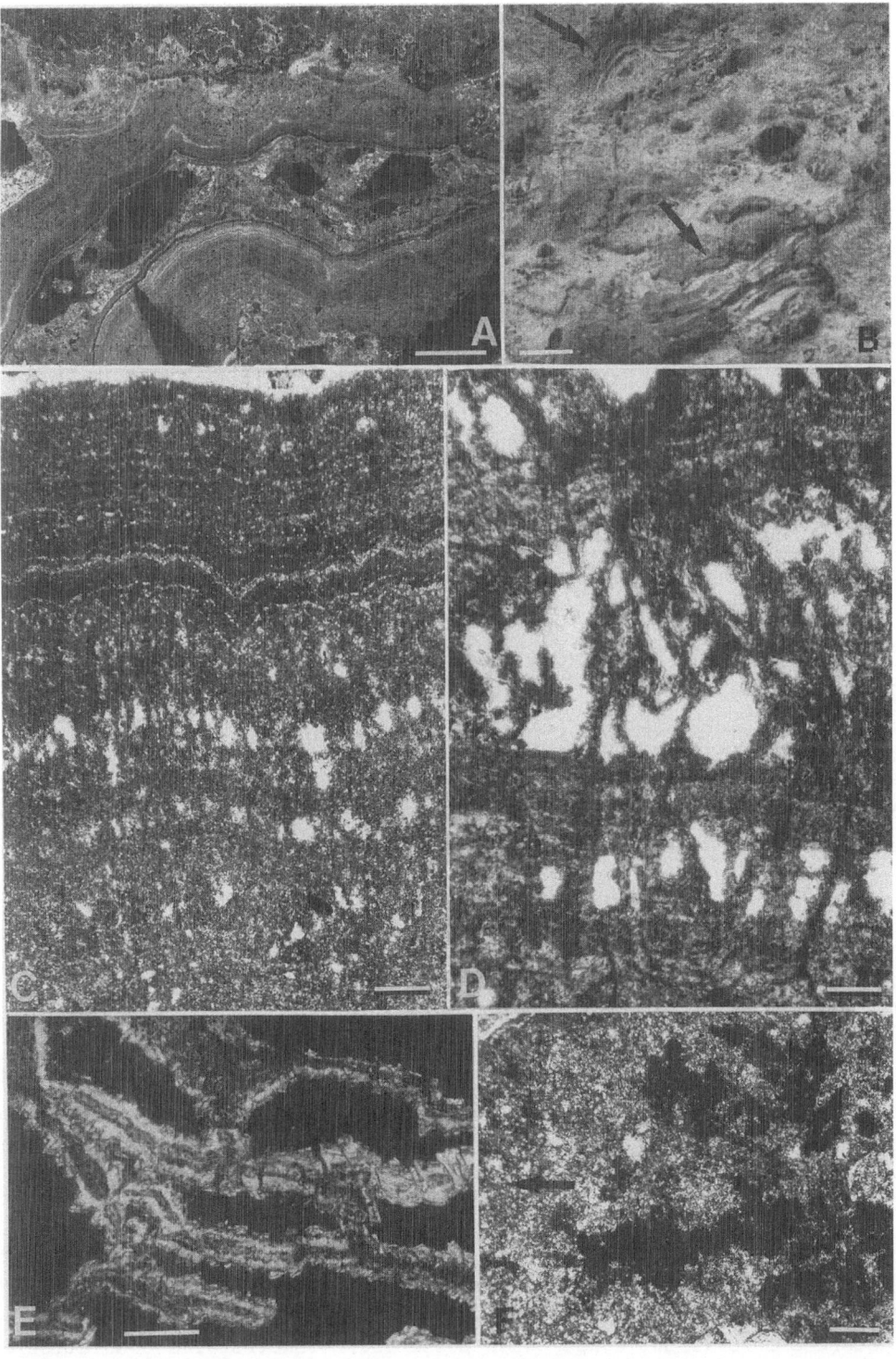

1982 a, 1984; VAN LAER and MONTY, 1984), these needles grow perpendicular to the substrate (carbonate host rock or internal sediment) or in juxtaposed fans. Late blocky calcites, drusy or not, may also, totally or partially cement these cavities.

Other cementation patterns may result, from the development of centimetric to decametric cavity dwelling stromatolites (endo-stromatolites of MONTY, 1982 b; MONTY and MAURIN, 1982; MONTY, 1984, 1986 b) such as figured in Plate 4, D, and Plate 5. These dolomitic biocements are initiated by the growth of microbial colonies of filaments supporting each euhedral and optically oriented dolomite crystals. These microbial growths and related cementation processes may be loose (Plate 5, D, E) or compact (Plate 5, A, C). Finally anhedral, or globular microbial (CRCC) dolomite (MONTY, 1986 a) may also cement laminoid or vuggy fenestrae as will be seen in the stromatolite section. Other aspects of replaced laminated sulfates are illustrated on Plate 2, C, D, and Plate 14.

Beside the small cavities described here above, decimetric to metric ones can also be found (Sharm el Bahari, Zeit, etc...) in different stages of infilling by internal sediments (Plates 4, D; 19, C) or cements; among these however, larger cavities may result from other processes than simple carbonatation of sulfates (salt movements and eventual dissolution, sliding, disruption of unsupported calcareous masses due to changes in volume in surrounding rocks, etc...)

Locally, these diagenetic carbonates are associated with sulfur bodies (Gemsa). SELLWOOD and NETHERWOOD (1984) indicate the presence of kerogenic laminated dolomitic "claystones" associated with sulfur from cores recovered from evaporitic cycles.

III. THE MARGINAL CARBONATE COMPLEX.

The growth features of the marginal carbonate complexes have been controlled by the geometry of the basement highs such as the Esh el Mellaha Range, the Gebel Zeit, the Coastal Ranges of the Red Sea, etc..., overhanging lows filled up with the evaporitic suites. The topography, the relative mobility and the presence of re-activation phases in the basement blocks had a major influence over the distribution, the thickness and facies of the different sedimentary sequences : coral build-ups when preserved, biodetrital accumulations,

145

GEBEL ESH MELLAHA CENTER

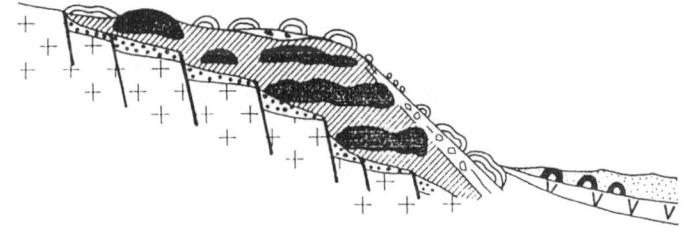

ABU SHAAR EL QIBLI CENTER

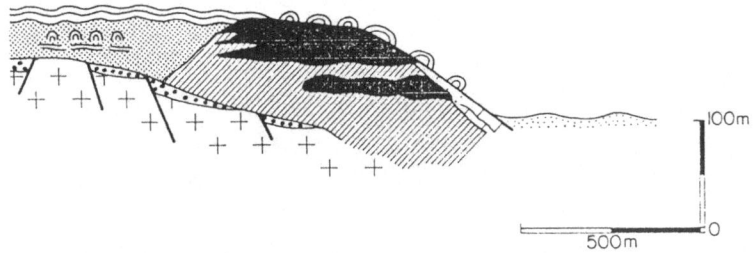

100m

500m

0

ABU SHAAR EL QIBLI SOUTH (WADI KHARAZA)

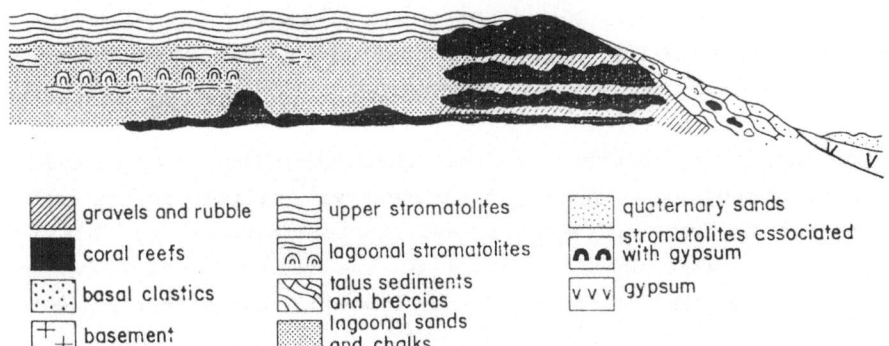

▨ gravels and rubble	⊜ upper stromatolites	⠿ quaternary sands
■ coral reefs	⌓⌓ lagoonal stromatolites	∩∩ stromatolites associated with gypsum
⠿ basal clastics	⧅ talus sediments and breccias	v v v gypsum
+ basement	▦ lagoonal sands and chalks	

Figure 2. Composite sections of three transects through the Abu Shaar el Qibli-Esh el Mellaha carbonate complexes. Location of sections is shown by empty arrows on the simplified map of Plate 1.

lagoonal deposits, thickness of talus, size of breccia components, erosional phases, etc... . In the northern part of the Esh el Mellaha, the seaward slope exhibits spectacular talus comprised of carbonate debris and siliciclastics (Plate 7,A). Sedimentary dips may be over 40°. Residual coral bodies of limited extension appear on top or edges of the complex. The most striking structures lie on the Abu Shaar el Qibli Plateau wrapping the southern tip of the Esh el Mellaha Range (Plate 1; Figure 2).

A. THE ABU SHAAR EL QIBLI REEF COMPLEX.

Excellent exposures and the presence of simple block faulting facilitate the recognition of the different phases of the carbonate complex : thick and sometimes broad talus, coral bioherms, lagoonal deposits including pinnacle-reefs, and shoals supporting small stromatolites (Plate 6). These main facietal units as well as the overlying stromatolitic cap have been previously, overviewed and illustrated by the authors (ROUCHY, 1979, 1982; ROUCHY *et al*, 1982, 1983) as well as by EL HADDAD *et al* (1983/1984) who added the presence of siliciclastic influxes disgarding previous publications and their illustrations.

The surface of the plateau, dipping smoothly to the south stands some 150 meters above the Quaternary plain where the uppermost part of the evaporitic formation, thickly deposited in the adjacent graben, pinches out against the base of the reef complex (Plate 6,A; Figure 2).

From north to south, the carbonate complex offers dominant calcarenitic sediments above transgressive siliciclastics. These calcarenites thicken southwardly and harbor nice coral build-ups sitting on the outer shelf in front of the trough (future evaporite basin). The thickness of the bioherms depends on the morphology of

Plate 6. The carbonate complexes of Esh el Mellaha-Abu Shaar el Qibli. **A.-** Aerial view of the reef front showing coral bioherms (a), eventually interspersed with detritus; the rubble layers and steeply dipping talus (b); the uppermost well bedded deposits (c) correspond to the "upper stromatolites" capping the complex and pinching out near the reef rim. Downslope, white gypsum beds (d) overlie talus carbonates. Approximate height of the talus is 150 m. **B.-** View of well bedded lagoonal sediments (c) topped by stromatolites; these sediments surround disconnected patch reefs (b) and interfing seaward with the main coral reef bioherms (a).

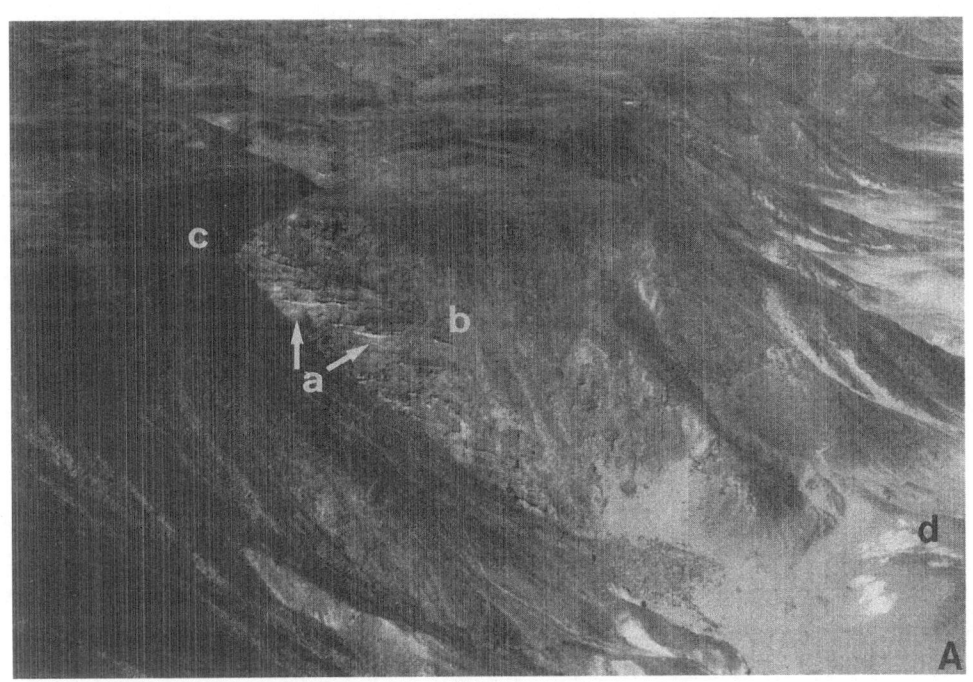

substrate and the tectonic history (including relative vertical movement of SW-NE faults); discrete in northern part, reefs develop southwardly to reach 60 meters in thickness along a narrow hectometric belt well exposed in two deeply cut wadis ("south Palmers" and Kaharaza). We accordingly disagree with EL HADDAD *et al* (1983/1984) who only reported bioaccumulated skelettal debris and siliciclastics and minor bioherms. Our observations reveal the presence of true coral reefs, more developed in the southeastern part, where they may form a true barrier eventually cut by "passes" probably related to transverse tectonics. This belt comprises coral bioherms pinching out lagoonward and eventually prograding basinward over strongly dipping reef breccia. They are built by *Porites* in association with a diversity of massive corals (faviids, astreids, etc... see GREGORY 1906), invertebrates (including reef borers) and rhodophytic crusts and nodules.

Behind this reef edge, a broad lagoon develops (Plate 6,B; Figure 2); sediment comprises a variety of bioclastic or lumachellic carbonates (bivalves, gastropods...) intermixed with some terrigenous sediments. Red algae and rhodolites may be abundant. In Wadi Kaharaza, lagoonal beds cover a coral sole and bear scattered coral pinnacles (Figure. 2, Plate 6.B).

The outer slope has a typical talus (Figure. 2; Plates 6,A; 7,D). Reaching several decameters in thickness in its southern part, this talus is volumetrically a very important part of the overall reef-complex. Sedimentary dips may locally reach 40° (ROUCHY, 1979, 1982; ROUCHY *et al*, 1983). This fore reef is made up of breccias (Plate 7, B, D) with blocks of coral patches, bioclastic carbonates and siliciclastic intercalations.

At places, the slopes smooth out a litle bit into narrow and discontinuous sigmoidal platforms; these appears to support (eventually) stacked bulges of thrombolitic to laminated dolomite (Plate 7,C) that will be detailed in the stromatolite section ("upper stromatolites"). Large sedimentary deformations have also been observed

Plate 7. A.- View of well preserved talus, northern Esh el Mellaha. Approximate height of talus is 150 meters. **B.**- View of strongly dipping beds of talus breccia (b) by-passing *Porites* colonies in living position (a) at the outer edge of a bioherm. Abu Shaar el Qibli. Scale bar is 50 cm. **C.**- Dolomitic, laminated or thrombolitic metric bulges sitting on sigmoidal flattened parts of the slope. Abu Shaar el Qibli, "south Palmers". **D.**- Close up view of the base of the talus. Metric wedges are highly porous and contain thin lenticular channelling beds of sandstones, clearly appearing in whitish hues, in between the uppermost 2 wedges.

on the slope.

The evaporites dipping 10° to 20° to the East crop out in the
Quaternary plain about 150 meters below the edge of the reef; they
pinch out against the talus (Figure 2; Plate 6,A), a fact which
postdates their deposition with respect to the development of reefs.

B. THE ABU SHAAR STROMATOLITES.

In this general framework, stromatolites and stromatolitic-like
build ups have been found capping all the parts of the complex (reef,
lagoonal deposits, talus) and intercalated within lagoonal sediments
(ROUCHY, 1979, 1982; ROUCHY et al, 1983) (Figure 2; Plate 6,A); on the
basis of paleogeographic locations, various morphologies and
microstructural types of microbial accretions can be recognized.

1. The lagoonal stromatolite.

Two contrasting types of lagoonal stromatolitic deposists have
been observed.

The first assemblage (Plate 8), is found near the base of the
lagoonal deposits in Wadi Kaharaza (Figure 2); it varies in thickness
from 50 cm. to 1 meter at least, as a result of strong erosionnal
processes at the top of the bed and/or heavy bioturbation.

Plate 8.- Abu Shaar El Qibli (Wadi Kaharaza) : lagoonal stromatolites.
A.- Complete sequence of stromatolitic bed : starting with binded sands
and slightly undulating mats (a); followed by closely stacked discrete
stromatolites (b) overgrown in their turn by columnar stromatolites
(c); this sequence is abruptedly interrupted by a discontinuity surface
overlain by bioturbated sediments (d). - Scale bar : 10 cm. B.-
Incomplete sequence showing the basal undulatory mats (a) directly
overlain by poorly developed and sparcely distributed columns (c) in
lagoonal sands and silts while the overlying bioturbated sediments (c)
have been severely eroded (arrow). Scale bar : 5 cm. C.- Complex
sequence in which bioturbation (d) occurred in several superposed
phases separated by discontinuities (crusts) marked by arrows. Scale
bar : 10 cm. D.- Strongly bioturbated stromatolites of phase (b)
showing their initial soft state. Scale bar: 5 cm. E.- General view
showing basal lagoonal sands and gravels overlain by stromatolitic
phase (a) : binded sand at the base passing to undulatory
stromatolites; initiation of discrete stromatolites (b); their growth
is interrupted by lagoonal fine sediments with synsedimentary
deformations underlined by white kinked discontinuities; bioturbated
horizon, (d).

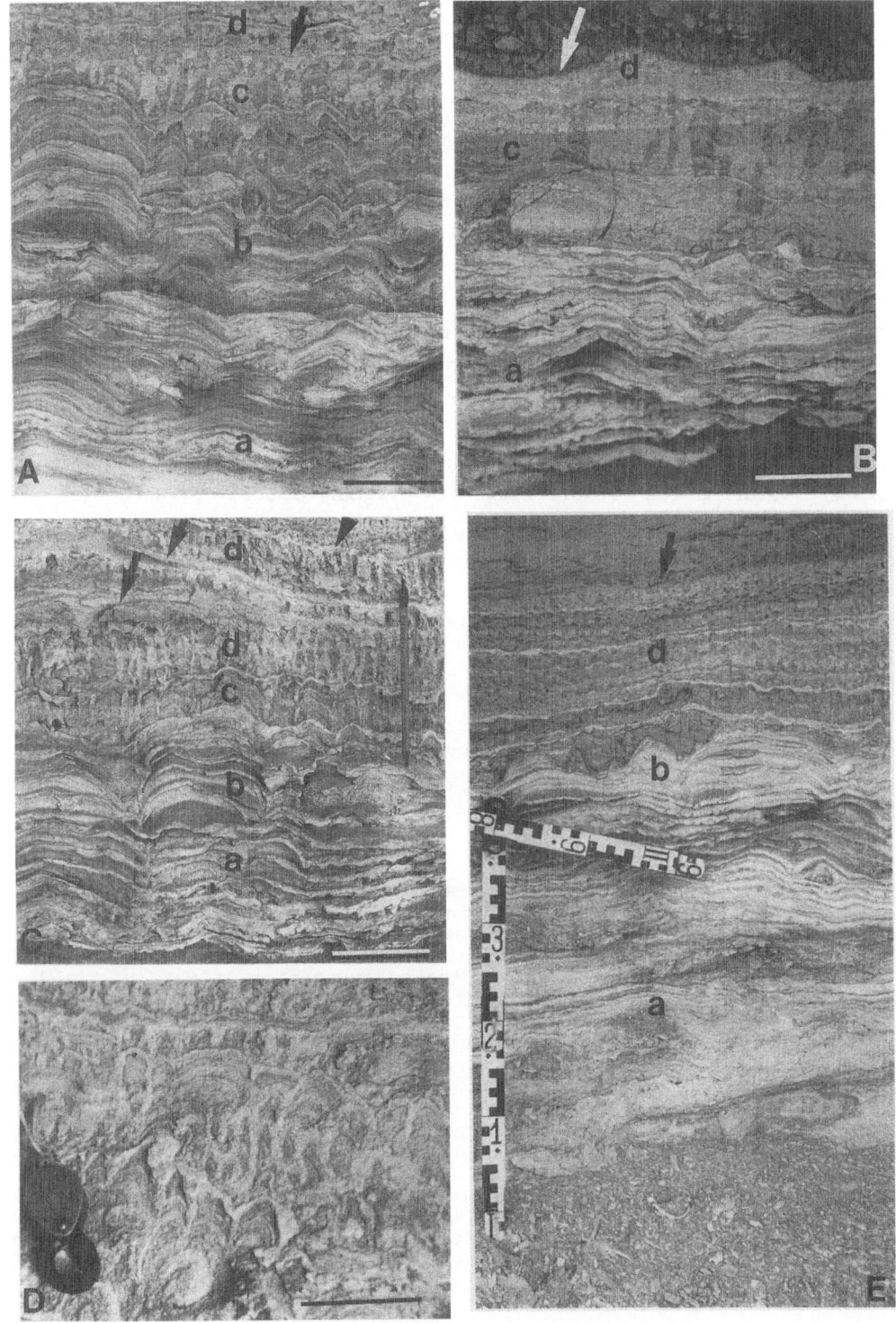

It developes on bioclastic gravelly sands (Plate 8,E), presents up section a succession of stromatolitic morphologies, to be finally interrupted by discontinuous erosive, millimetric to decimetric deposits of lagoonal yellowish green sands overlain by a coquina.

The assemblage begins with planar passing to slightly undulating mats (stage (a) on plate 8). It shows a lamination resulting from alternating layers (about 500 μm thick) of equigranular, globular and coarse dolomite, with layers or laminae of microsparitic dolomite. Locally this lamination appears bioturbated and replaced by pockets of elongated faecal pellets (maximum 70 μm x 350 μm); at places these may form individual laminae; these mats, separated by layers of binded lagoonal sands and silts have accordingly been heavily browsed by unfossilized animals. They furthermore show remains of kerogen and clear orange organic matter.

These planar to wavy microbial mats, which probably represent the pioner colonisation stabilizing the lagoon floor, pass upward to discrete contiguous stromatolites, 5 to 10 cm. high (phase (b) on photos A,C,E of Plate 8). They show a complex microstructure that we shall briefly describe here by considering the microstructure of the area framed in Plate 9,A and enlarged in Plate 9,C, area which is representative of the stromatolite constitution. It shows three orders of laminations. The first order lamination comprises a succession of three microstructural types of layers, each a few mm. thick, (cyclothemic lamination of MONTY 1976); this succession is repeated "up section" in the stromatolite with eventual slight variations in developments and minor features.

Plate 9. A.- General view of stromatolite from phase (b) on Plate 8. (negative photograph). The area framed in white is detailled in C to show the complex organization of these stromatolites. Scale bar : 5 mm. **B.-** Detail of A. (negative photograph) showing the three basic types of first order lamination (1, 2, 3), the superposition of which forms the basic microstructure of the stromatolite. The first layer (1) consists of a fenestral (microbial) dolomite; it is poorly to non laminated and may show ghost of vertical filaments. The second layer (2) displays a second order lamination resulting from the succession of dense dolomicritic to dolomicrosparitic laminae (a) and porous fenestral laminae (b); both of the laminae contain ghosts of vertical micritic filaments (horizontal arrows in layer 2); the micritic laminae on their side may have furthermore preserved thin tightly packed micritic films (vertical arrows at the base of layer 2.). Scale bar : 1 mm. **C.-** (negative phtograph) View of bioturbated zone (phase d on plate 8) showing bioturbated sediment with vertical fenestrae (traces of reeds ?) abruptedly capped by a dense layer of locally laminated dolomicrite forming the white layer in Plate 8 C (arrows) and Plate 8 E (zone d). Scale bar : 5 mm.

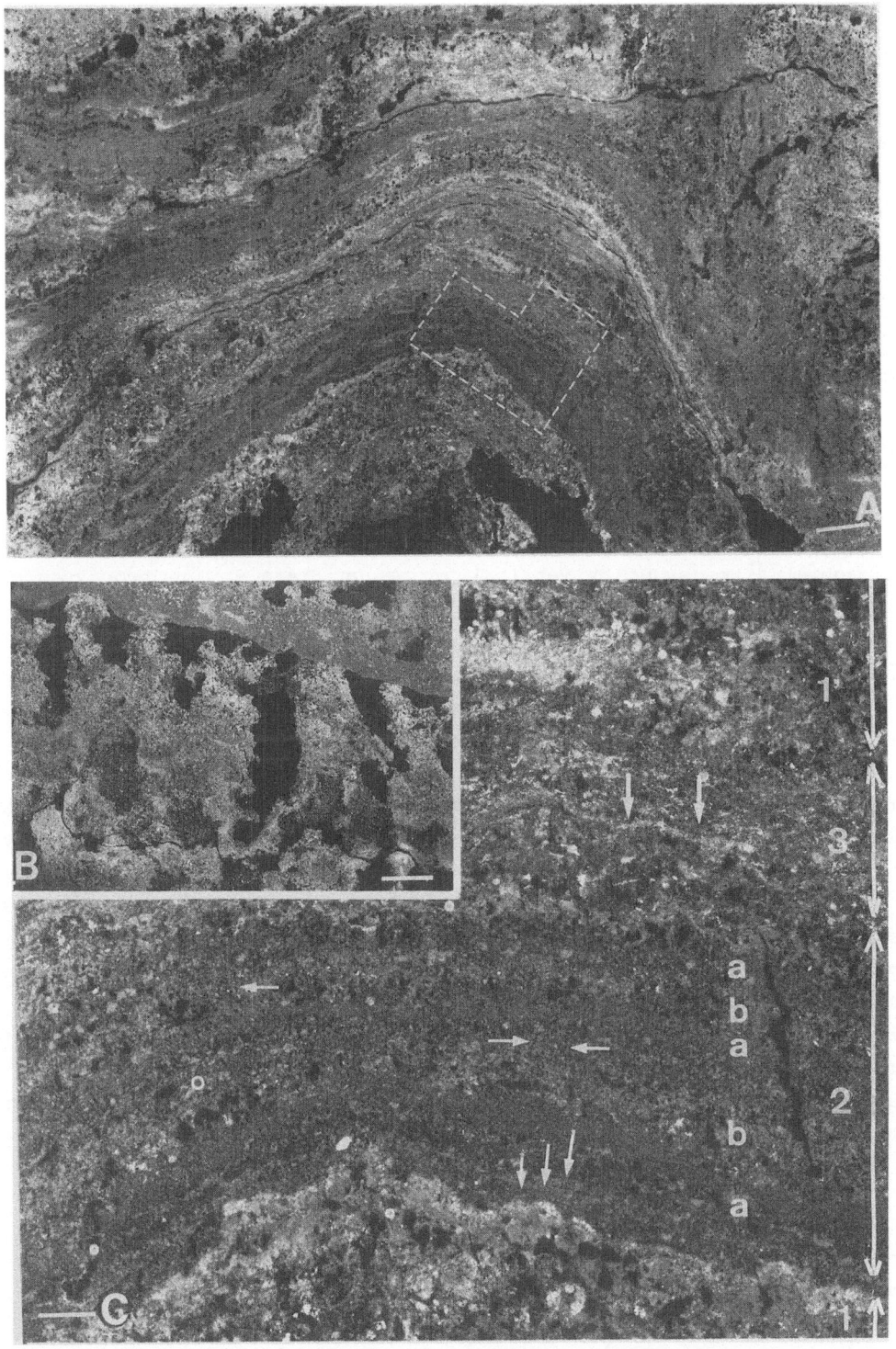

The first layer (1 and 1' on Plate 9,C), which starts stromatolite growth, is made of a fenestral microsparitic dolomite. Fenestrae are generally smaller than 1 mm, although larger ones can be found often vertically elongated and irregular in outline. The dolomite is rich in microbial intraformational rounded peloids (less than 30 μm in diameter) made of micritic-like dolomite (see below). This layer may show a vague lamination and traces of erected micritic tubes (diameter lower than 20 μm).

The second layer (2 on Plate 9,C) is more complex and shows a second order lamination of the type "simple alternating lamination" (MONTY, 1976). This lamination comprises a regular alternation of micritic to finely microsparitic laminae (up to 500 μm thick, (a) on Plate 9,C), eventually rich in scattered clusters of irregular to globular microbial dolomite showing a central black dot (see below), with more porous laminae ((b) on Plate 9,C) rich in intraformational peloids and small fenestrae. At places, the micritic laminae have preserved a third order lamination made of alternating clear and dark microbial films shown by vertical arrows on Plate 9. The two types of laminae show remains of vertically growing micritic filaments. On the sloping sides of the stromatolite this second layer presents what appear to be tension cracks which do not pass into adjacent layers (Plate 9,A; 9,C); this may reveal an original soft or gelatinous state of mats mostly made of vertical builders (compare with third layer).

The third layer (3 on Plate 9,C) is compact and brownish in reflected light; it shows an abundance of horizontal films either rich in light orange organic matter or dolomitized (arrow). The dolomite constituing this layer is also rich in intercrystalline and intracrystalline translucid organic matter; one also notes the abundance of brownish entangled filaments as well as of filament debris after collapse of supporting sheath organic components; these debris of filaments may look like peloids. Beside the presence of organic matter, brownish color of this layer is enhanced by the presence of patches of iron oxides probably after sulfides. This third layer built by compact and prostrate filaments (horizontal films) pervading and binding erect filaments must have been very dense and tight; it may recall the *Lyngbya-Microcoleus* association.

These discrete stromatolitic domes and heads are in their turn overgrown by discrete, elongated to columnar forms, up to 10 cm high, clearly separated from each other (Plate 8,A,C, phases (c)); laterally (Plate 8,B, about 100 m basinward of Plate 8,A) these columns may becomesparsely distributed in yellowish lagoonal silts and sands. Note that in all cases their growth starts from a thin white layer of binded sands underlining discontinuities. The microstructure of these columns is similar to that of the stromatolite described here above.

At this stage, stromatolite growth become completely disturbed by bioturbations (Plate 8, A, C, E, phase d; Plate 8, D) or erosional processes (Plate 8, B). In some cases bioturbational processes start early and stress the developments of stromatolite (b on Plate 8, E); in other situations "burrow-like" bioturbations are distributed into superposed horizons separated by white indurated layers (arrows on Plate 8, C); thin section in one of these horizons is figured on Plate 9, B. One can see a compact micritic layer on top of photograph, capping the underlying "pseudocolumnar" bioturbated sediment. This layer is made of fine almost equigranular dolomite with rare fenestrae; the clusters of peloids scattered in the fine dolomite, and the undulation of the layer over a protruding pseudo-column in the upper left corner could indicate that this layer could have been initially a rather soft mat that rapidly indurated into a crust. This is justified by the fact that at places, this layer is clearly laminated and made of alternating dense dolomitic laminae (around 1 mm thick) and laminoid fenestral laminae as will be commonly seen in overlaying stromatolitic horizons. The underlying pseudocolumnar part is a highly bioturbated dolomite, as can be seen from the mixture of lithologies and the presence of disrupted, displaced pieces of mats (white undulose streaks). The shape of the intervening vertically elongated cavities – if original – would evoke traces of plants like reeds as their size is much to small for mangrove roots or *Thalassia* rhizomes or yet *Halimeda* holdfasts, the more that there is no vestige of *Halimeda* around; these fenestrae compare very well with traces of living and Messinian reedfield found in association with stromatolites around the Santa Pola reef (Province of Alicante, Spain). These cavities may be partly filled with loose sediment or cemented by sparry calcite or gypsum (with anhydrite relics). Whether these sulfates are precursors of the evaporitic phase or remobilized from miocene sulfates is not always clear.

The sharp contact with coquina and normal lagoonal sands, the discontinuities and erosional features (see introduction of this section and Plate 8, B) suggest that these stromatolites may have developed on temporary and periodically flooded shoals or flats in the lagoon. Finally, the fact that stromatolites themselves (phase b on Plate 8) have also been bioturbated by unidentified organisms, confirms their initial unlithified nature; lithification most probably resulted from dolomitisation processes.

In "South Palmers" area, another type of bioconstructions has been found in the lagoon at about 2-3 m above the coral sole (Figure 2). These cavernous, stromatolitic-like rocks appear very complex on outcrop as can still be appreciated on polished slab (Plate 10,A). The base of the slab shows slightly undulating stromatolites (a) overlain by a whitish, compact and vertically cracked irregular layer (b) of compact microsparitic dolomite containing small quartz grains; this dolomite has the same texture as that forming the whitish discontinuities on Plate 8,C and Plate 9,B. At places, this layer appears as injections in the surrounding sediments or in fissures. This lower part is overlain by a complex mixture of stromatolitic mats and calcareous grains. This is illustrated on Plate 10,B showing a representative area located below the vertical arrow on Plate 10,A. The main sediment is here a sandy coquina (molluscs, gastropods, serpulids, forams) (b on Plate 10,B) where moldic shells have been replaced by dolomite. This grainstone contains small rhodolites (such as pointed by horizontal arrow on plate A) and complex irregular nodules (r on Plate 10,B) either almost completely built by melobesians (nodule to the right), or consisting of complex mixture of dolomitic mud, quartz splinters, fossil molds, serpulids, pellets, and discontinuous thin plates of red algae (nodule to the left); the surface of this nodule (marked by small arrows) is almost completely delimited by platy melobesians. The upper part of Plate 10,B shows a dolomitized microbial mat capping bioclastic sand (this mat is indicated by a vertical arrow on Plate 10,A). Lamination of this stromatolitic mat results from alternation of dense, discontinuous dolomitic films or laminae, and

Plate 10. A.- General view of lagoonal dolomitic stromatolitic rock (Wadi by "South Palmers") a few meters above coral reef sole (Figure 2). Note sinuous stromatolite at the base (a) overlain by dense and fine dolomicrite (b). The overlying part is a complex mixture of small stromatolitic developments, bioclastic sands, small rhodolites (horizontal arrows) and complex rhodophytic nodules. Vertical arrow points to top of a microbial mat visible in the upper part of photo B (Ph. C.M.). **B.-** (negative photograph) Detail of A (below vertical arrow). It shows an irregular rhodolite (r) to the right and a complex rhodophytic nodule (to the left, periphery of nodule is marked by small arrows) in a lagoonal skeletal sand with very high moldic porosity (b). On top, lies a stromatolitic mat made of dense dolomicritic laminae (whitish) alternating with porous laminae made of erect dolomitized filaments. Scale bar : 2 mm (Ph. C.M.).

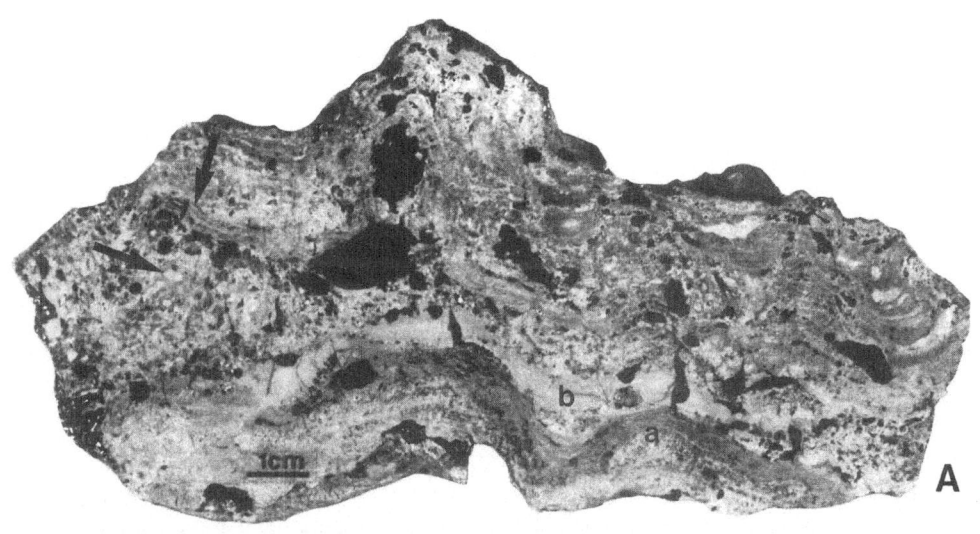

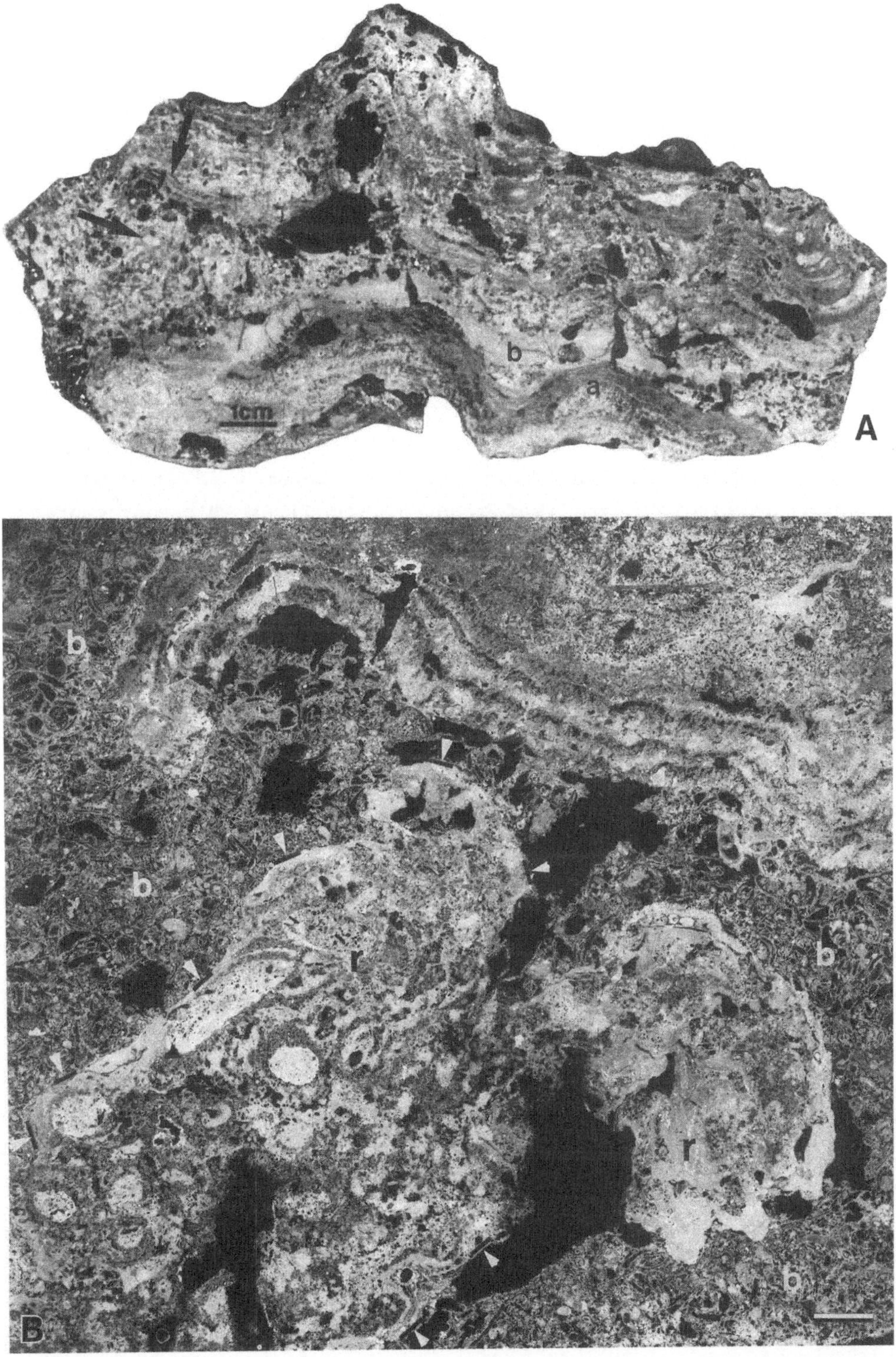

A

B

looses, porous laminae originally made of erect filaments, now
overgrown by fine dolomite (process similar to that illustrated on
plate 5,E). The overall observed association indicates that these
stromatolites grew in normal marine lagoonal waters.

The second lagoonal stromatolitic horizon (Wadi Kaharaza) about 50 cm.
thick, is found a little bit higher in the section than the first one
(Figure 2). Squeezed between two beds of lagoonal bioclastic sands
(Plate 11,B), it already shares many common features with the "upper
stromatolites" (see below) as well as with the Messinian stromatolites
capping the Santa Pola coral reef (Spain). They might thus record a
first limited pulse preluding the installation of the true evaporitic
phase. These stromatolites, planar to slightly and regularly undulatory
when observed across the strike (Plate 11,B, offshore is to the left)
are discontinuous and form flat elongated domes when observed in a
direction parallel to the reef front (Plate 11,A). Microstructure is
quite variable both vertically and laterally. Briefly it shows an
alternation of dense "micritic" to microsparitic dolomite (white on
Plate 11,C) with loose fenestral laminae; as in many previous cases,
"micritic" dolomite refers to microsparitic to sparitic crystals which
look micritic as a result of the tremendous amount of inclusions and/or
abundance of nannocavities left by oxydized bacterial remains (in this
case SEM observation shows highly fenestrate crystals almost like

Plate11. Second level of lagoonal stromatolites, **A.**- View of
stromatolitic sample (see B) showing traces of reeds (arrow) under the
upper dense dolomitic layer. Wadi Kaharaza. **B.**- General view of second
stromatolite bed. **C.**- Microstructure of stromatolites showing
alternation of dense dolomitic laminae and vacuolar sometimes very
loose ones (negative photograph). In the uppermost part of the
photograph loose laminae show collapsed fragments of dolomitized
filaments (arrows). In the central part of photograph small or larger
residual cavities persists among dolomite developing centrifugally from
filaments; in lowermost part of photograph, loose laminae present an
arachnoid microstructure resulting from preserved and unbroken
dolomitized filaments, whereas interstitial cavities remained opened.
Scale bar : 2 mm. **D.**- Microstructure of stromatolite from the upper
blanket plastering the reef (negative photograph). It mostly reveals
four types of microstructures : (1) finely laminated on top where
alternate thin dolomitic films with porous layers; porosity is formed
by residual cavities in the dolomite developing centrifugally from
presently dolomitized erect filaments (2) highly porous generally
poorly laminated thick layers, rich in faecal pellets (white dots);
fenestrae may be ameboid or replicate the shape of vanished sulfates
(3) dense micritic layer (4) laminated mat made of alternating laminae
of microbial dolomite and open laminoid fenestrae. This mat is cut by
oblique, parallel and sigmoidal distensive cracks. Scale bar : 1 mm.

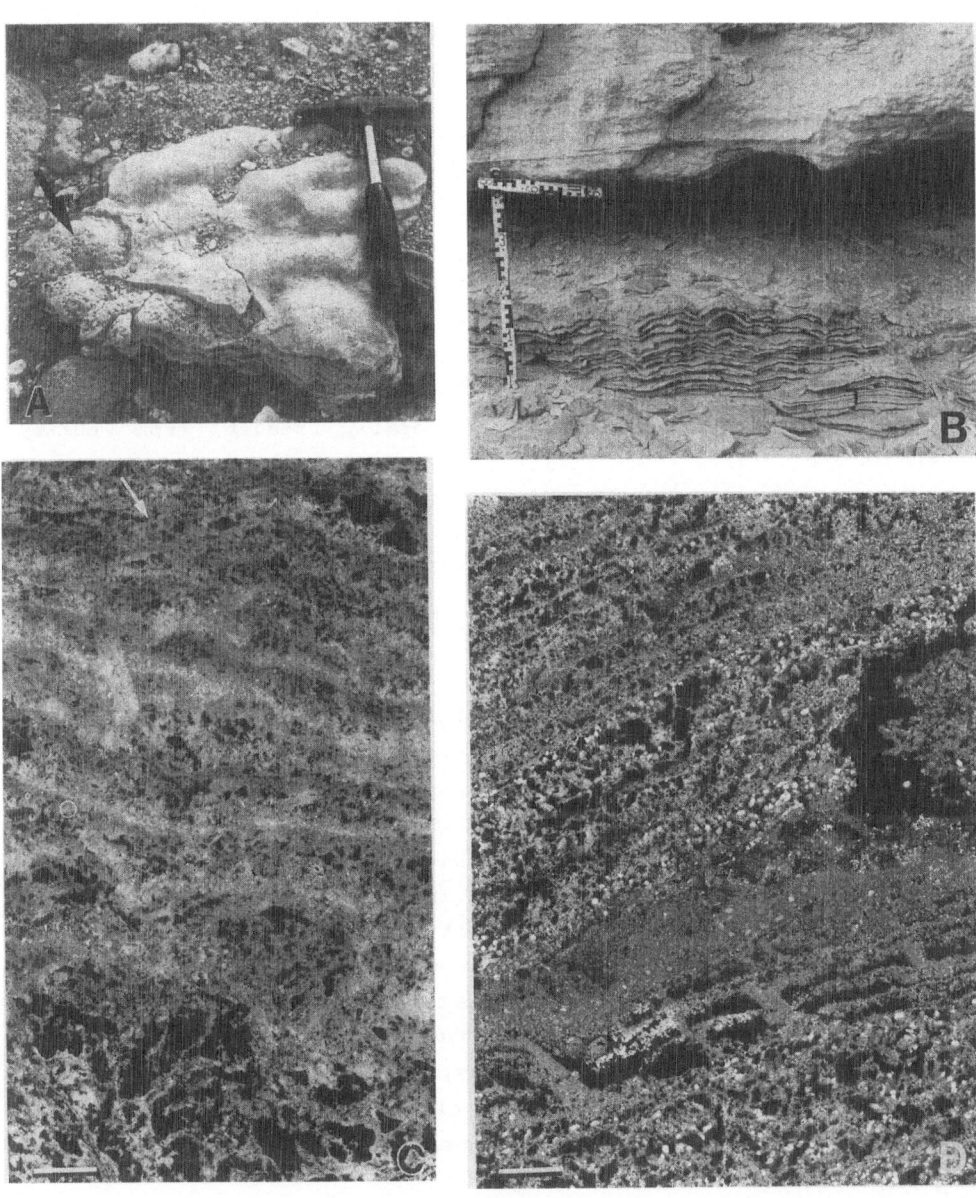

lace); when crystals are closely packed it is sometimes very difficult to trace their boundaries so that the resulting mosaic looks micritic. As for microsparitic to sparitic dolomite all the intermediates are found between globular dolomite, developed around a bacterial colony (such as on Plate 12,D) or a droplet of kerogen, to pseudo-hexagonal zoned dolomite and from there to zoned rhombs as explained in MONTY 1986 a. Furthermore, oxidation or, on the contrary carbonatation of organic matter may lead to hollow globules, crystal, or crystal zones (eventually in process of cementation) on the one hand, or to micritisation and formation of CRCC dolomite on the other (MONTY, 1986a).

The loose vacuolar layers generally result from outward development of dolomite precipitation into the interstices left between dolomitized filaments (see Plate 5) which serve as initial substrate. In some of these layers one can see fragments of such dolomitized filaments broken down into pieces (early compaction) and lying disorderly in the dolomite (Plate 11,C, white streaks in topmost part of negative photograph). In other loose layers (lowermost part of Plate 11,C) dolomitized microbial filaments design an arachnoid eventually bacinelloid pattern. Beside pseudo-peloids resulting from "micritized-like" dolomitic globules, true micritic peloid s.s. may be found in these layers.

It is yet interesting to note that sulfides are found within the dolomitic mosaic whereas sulfates may occlude cavities. Both are probably unrelated if sulfides result from bacterial sulfate reduction of SO_4^{2-} radicals of original interstitial waters or of organic matter closely associated with dolomite, whereas sulfates appear younger and bound directly or indirectly (reprecipitated) to the evaporitic event.

Finally it is worthwhile noting that these stromatolites have also been bioturbated, showing that they were initially soft before suffering strong diagenesis. Traces of bioturbation appear on Plate 11,A, where the arrow indicates the lumen of vertical cavities probably due to reed, and on Plate 9.

2. The upper stromatolites.

Lagoonal deposits, coral bodies and associated bioclastic sediments are covered with an upper well-bedded stromatolitic formation (Plate 6; Figure 2); stromatolitic beds are up to several meters in thickness, and extend over several kilometers (ROUCHY, 1979; ROUCHY et al, 1982). The formation up to 30 meters thick overrides the outer edge and talus slope with scattered algal heads (metric to decametric) associated with slumped stromatolites. In most situations, a discontinuity separates the stromatolitic formation from the underlying sediments.

Upper stromatolites display striking morphologies with large domal structures several meters wide and up to one meter high (Plate 12, A); the domes may be basically organized into smaller decimetric heads or columns. Undulating planar stromatolites are also found; some of them may be composed of manganese oxides.

Various types of microstructures can be recognized according to types of mats but also according to diagenesis; many of them however are reminiscent of microstructures described in the second level of lagoonal stromatolites.

a. A common microstructural type is illustrated on Plate 11, D (negative). One can see on top a laminated pattern resulting from the alternation of thin dolomitic films (70-100 μm) with thicker highly porous laminae (up to 500 μm); this porous pattern results from the presence of vertical or sinuous chains of dolomite crystals having overgrown microbial filaments (see above and Plate 12, B). This distinct lamination becomes progressively obliterated toward the upper left corner as result of continued growth of microbial dolomite in the fenestrae left between the former filaments. The middle and lower parts of photograph (Plate 11, D) present, highly vacuolar layers, faintly laminated in the former site, non laminated in the latter one; such layers are up to 1 cm thick or even more. Fenestrae may be irregular and ameboid, or replicate the outline of vanished sulfates. They may be rich in elongate pellets (about 70 μm X 350 μm) which appear as white dots on photo namely at the base of the central porous layer and scattered through the basel one. Laminoid cavities may be cemented by gypsum or large calcite crystals; these are late cements as they also occur in fractures. These two highly porous zones are separated by a laterally discontinuous layer of compact unlaminated fine dolomite; it overlies a laminated mat made of layers of globular microbial dolomite (see Plate 12) separated by uncemented laminoid fenestrae. This mat is cut by a series of oblique parallel and sigmoidal fractures filled with very fine grained dolomite; their pattern does not represent dessication cracks but rather distensive shear cracks. Note the way the sediment of the first cracks to the left intrudes the overlying massive dolomite, a feature that might evoke injection. In spite of all these diagenetic features, the presence of fecal pellets laid down in the

porous layers suggests intense browsing, hence an initial soft state of
these mats.

Another aspect of the laminated microstructure illustrated on
Plate 11, B is shown on Plate 12, B; the base of photograph shows a
lamination composed of thin micritic films separating porous laminae
with vertical dolomitized filaments (as on top of Plate 12, B); here
however, this microstructure passes upward to a vuggy dolomite, with
irregular, ameboid fenestrae, where faint denser horizontal laminae can
still be guessed.

b. Among the upper stromatolites, some show a very faint
lamination and appear extremely compact and homogeneous on outcrop;
only their domal morphology attracts the attention. Microscopic
observation also reveals their very complex constitution. The
microstructure is the more complex that its main features are very
discontinuous laterally and present various microfacies vertically.
Basically, one can recognize a faintly delimited alternation of dense
"micritic" dolomitic layers, laminae or bits of films, such as appears
in the middle of Plate 12, C, with sparitic layers of generally
globular microbial dolomite impregnated with yellowish organic matter
(Plate 12, E). Dense laminae result from at least four different
processes: (1) highly compacted and horizontally stacked (fragments of)
dolomitized filaments; (2) replacement and dolomicritization of

Plate 12. **A.**- Domal stromatolite belonging to the upper stromatolitic
blancket. Abu Shaar el Qibli. **B.**- Thin section (positive photograph)
showing another aspect of microstructure illustrated on the uppermost
part of Plate 11., D; here however the classical alternation of thin
dolomitic films with loose filamentous laminae (base of photo) passes
upwards to a vuggy dolomite with ameboid fenestrae and faint horizontal
laminae of dense micrite. Scale bar : 500 μm. **C.**- and **E.**-
Microstructure of very compact faintly laminated stromatolitic rock
(positive photographs). **C.**- Alternation of sparitic dolomitic layers,
with eventual sulfate pseudomorphs, with dense "micritic"-like laminae;
in this case micrite pattern results from the very high content of
dolomitic crystals in residual organic matter and very fine sulfides
(see text for discussion and description of other types of "micritic"
laminae). Note the presence of horizontal streaks of kerogenous films.
Scale bar : 100 μm. **D.**- Close up view of microbial dolomite. Central
part shows a dolomitic globule in process of differenciating straighter
outline with protruding ridges, some of which are rhombic. These ridges
may be guided by microbial filaments extending from the dark organic
core (out of focus here). To the left, two dolomitic crystals show
their inner part made of a tangled microbial colony. Scale 50 μm. **E.**-
Microstructure of very compact faintly laminated rock (positive
photograph, view of a sparitic layer of microbial dolomite, with
scattered eventually tilted bits of organic films. Note all the phases
between globular dolomite with central black dot of microbial colony to
sort of "micritized" anhedral to subhedral crystals heavily loaded with
organic remains. Scale bar : 100 μm.

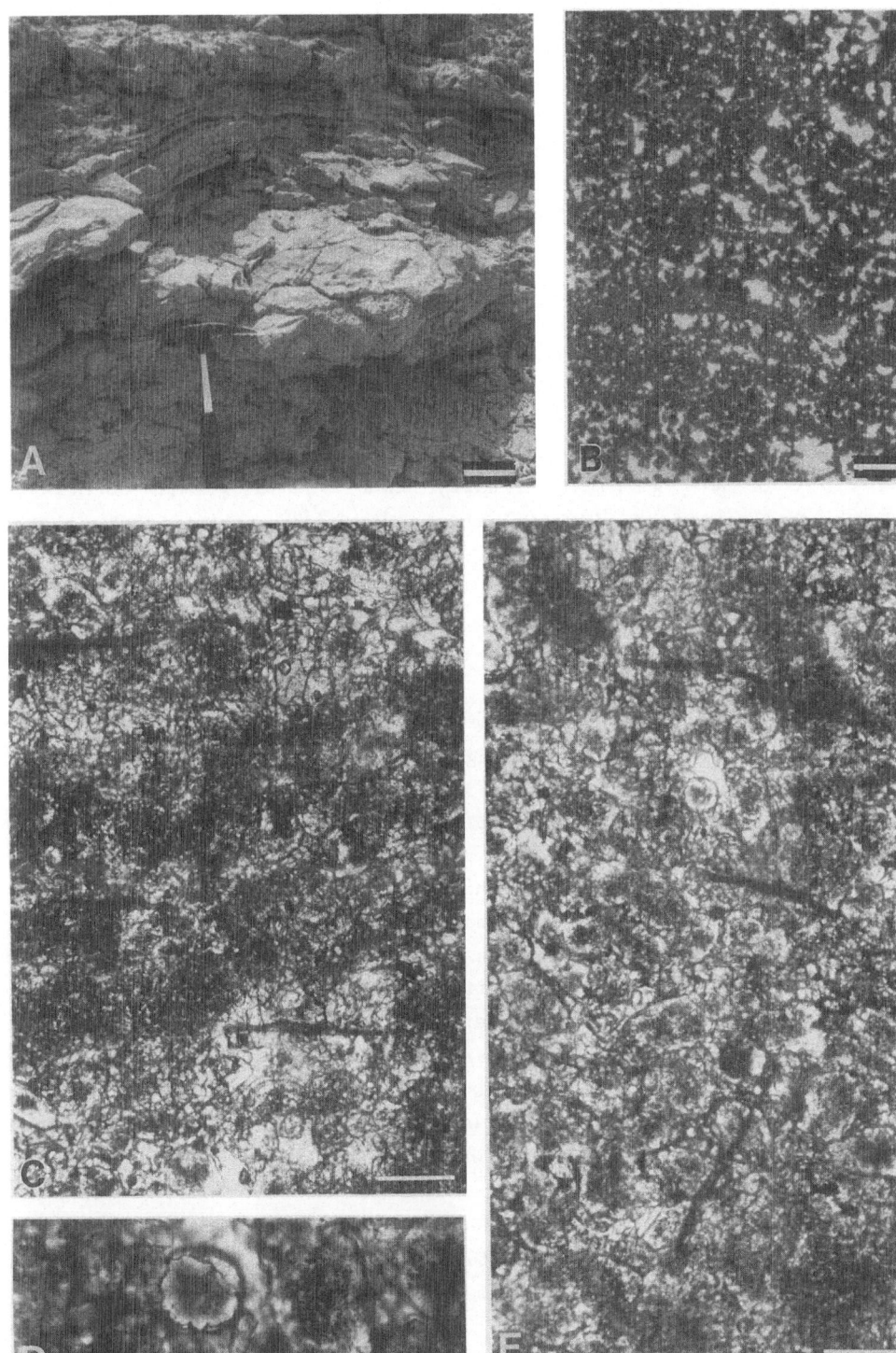

originally piled up organic films (some of which have been preserved as such and allow reported interpretation); (3) development of dense pseudo-peloidal layers due to micritic dolomitization of the organic core of dolomitic globules, core which may be very large with respect to the clear or less dirty dolomitic rim; (4) development of dolomitic crystals around microfilamentous microbial colonies (see crystal to the middle left of Plate 12, D) which may be limited in size and yield a so called black dot (CRCC) dolomite or occupy the entire crystals.

Observations suggest that growth pattern of microbes influences the crystal shape; when microbial colony forms well circumscribed small ball with eventual centrifugal filamentous projections, a globular crystal is iniated (Plate 12, D, center); when colony grows an elongated assemblage of filaments roughly oriented parallel to each other, an oblong crystal is formed (Plate 12, D, elongated crystal to the right of the globule). Finally when crystal growth constantly follows development of microbial colony, a dirty, "micritized" crystal is formed. When closely packed, such crystal yield a would-be "micritic" dolomite mosaic (see central layer of Plate 12, C and "micritic" patches of Plate 12 E). As said above such crystals may look like lace when observed in electron scanning microscope after complete degradation of original microbes.

Coarse sparitic layers (Plate 12, E) are made of heterogranular globular dolomitic crystals ranging from globules with a central black dot representing the initial microbial dolomite or, in some cases, a dolomitized kerogenous droplet, to hypidiotopic dirty crystals rich in organic inclusions as explained above. Careful study of black dot microbial dolomite reveals a whole series of forms leading from globules to rhombs : round original globules centered on a microbial colony first develop straighter outline with protruding small rhombic ridges eventually guided by microfilamentous projections from the core (Plate 12, D); such crystals, eventually coated by another bacterial organic film, progressively develop a (pseudo-?) hexagonal outline; modifications in the relative growth rate of faces initiate rhombic outline. This process may end up with zoned dolomitic rhombs centered on a black dot, the limits of which may be clear cut or diffuse. Successive organic coating may be altered to dolomicrite, like the core, or may be completely degraded (by oxidation of organic matter). This leads to rhombs with completely or partially hollow center, and hollow zones. These microcavities may be later cemented by various carbonate (calcite or dolomites, ferroan or not, according to the amount of residual organic matter, acting as a reducing agent). Such sequences have also been observed in dolomites from the Quseir area, as well as in mediterranean messinian dolomites.

Finally sparitic layers, although compact, present (locally) significant intercrystalline porosity due to the globular shape of the dolomite; some of these cavities are filled with kerogenous orange and translucid organic matter. These organic patches also appear as privileged sites for nucleation of dolomite which grows as swarms of small crystals, to finally form microsparitic dolomitic mosaics, here also all the developmental phases can be followed. This process accounts for the presence of patches of dolomicrosparite scattered among the sparitic mosaic, or for the presence of heterogranular dolomitic layers showing sparitic rhombs intermixed with, or cemented by, microsparitic dolomite.

One should yet note the presence in these stromatolites of scattered replaced sulfates as well as of sulfides.

c. The last microstructural type that will be described in the upper stromatolites is characterized by their thin regular lamination (from 100-300 μm) which presents two facies according to the studied zone across the stromatolite. The first one shows the now classical alternation of thin and discontinous dolomitic films (a few μm thick) separating (micro) sparitic layers of microbial dolomite; small crystals or isolated mosaics of heterotopic crystals of quartz with rolling extinctions are found scattered into the microbial dolomite; this dolomitic mosaic shows a relatively high porosity partly due to the former presence of sulfates (gypsum and anhydrite molds). In other place laminae of carbonate (100-300 μm) clearly alternate with laminae of small (15-30 μm) heterotopic quartz crystals. Carbonate pseudo-morphosed gypsum may also occur in patches or scattered in laminae. These stromatolites are characterized by a strong diagenesis affecting sulfate and siliceous phases.

3. The upper stromatolites : reef front and talus.

Various types of what appears to be microbial rock bodies are found on the slope .

a. The most obvious ones are undulating laminated dolomitic deposits lying into a confused, highly diagenetically altered dolomite with eventual clasts and debris (Plate 13, A). Closer look at these deposits shows that they most probably represent slumped stromatolites or laminites; some of them, very limited in extent, and presenting abrupt terminations constitue torn away stromatolitic pieces detached from their original slumping mat.

b. Metric to decametric build ups, such as illustrated on Plate 7, C, may be found on the slope itself or on small discontinuous terraces. They may appear cavernous, irregularly fenestral, massive or finely laminated; this lamination is tentatively illustrated on Plate 13, B (arrows). These dolomitic bulges may lie on the talus or may be overlain by a flow of reef breccia. Plate 13, B shows the discontinuity of the lamination which disappears in the central massive part of the outcrop. Such situations result from various intensities of dolomitization and diagenesis including sulfate reduction.

Fenestral to cavernous masses of rock seem to constitute an important part of these build ups. Their microstructure and petrography is very complex. We shall briefly describe a representative example where some sort of lamination can be seen; the reported succession and its main features (size, porosity, definition of laminae and their regularity) may be highly variable laterally. The rock comprizes the following elements : (1) very irregular layers of coarse sparitic calcite containing irregular rings of kerogen (up to 1 mm in diameter) such as found around aragonitic fans in biodiagenetic carbonates of Quseir; traces of acicular crystals, eventually pseudomorphosed in calcite, are found here in the calcitic mosaic delimited by the kerogen-rich rinds. The sparitic calcite contains centimetric cavities partially cemented by gypsum. This whole system is very reminiscent of cementation of laminoid cavities after sulfate reduction in presence of organic matter. As we shall see, the actual gypsum and calcite are late cements; (2) laminae (about 150 μm thick) made of closely spaced clusters of strontianite forming fans, or radially growing from a common center. Residual cavities are filled with calcite; (3) sparitic to microsparitic layers of slightly dusty anhedral dolomite. This coarse dolomite contains seams of kerogen (10-30 μm thick) closely superposed at places, these films, originally limpid, may become

Plate 13. A.- Slumped stromatolites or laminites capping the slope. They are embedded into a very heterogeneous dolomite with eventual clasts. Abu Shaar el Qibli, "South Palmers". Scale 15 cm. B.- Close up of a bulge or build up such as illustrated on Plate 7, C. Note the fine discontinuous lamination indicated by arrows. This lamination is very discontinuous laterally due to intensity of dolomitization and cementation after microbial sulfate reduction.

charged with small dolomite crystals while the organics darken (see Plate 12); isopachous dolomitization around the films increases their thickness up to 50-100 μm. Compaction, or volume changes due to diagenetic processes, lead to the disruption of these mineralized films into fragments which appear either scattered and oriented in all directions (such as in plate 12, E but with a much higher concentration of fragments and a stronger mineralization), or tightly packed horizontally; in this latter case they form a dense dolomitic lamina with calcite in residual porosities; (4) such laminae may be followed by another lamina of strontianite to which succeeds a new dolomitic layer as above; (5) one also finds discontinuous layers or pockets where mineralized fragments of films float disorderly into a sparry calcite mosaic. This and similar situations reported above (although in dolomitic matrix), may be understood by considering calcitized laminated sulfates such as figured on Plate 14, C; here, one can see well preserved orange-brown kerogenous films into a calcitic matrix (Zeit, ROUCHY et al 1985); from there, and considering the existence of fragments of films in the central calcitic mosaic of Plate 14, E, one can easily imagine the collapse of our brittle dolomitized films (with intercrystalline residual bituminous matter) into porosities left after sulfate dissolution or replacement, and their cementation by calcite. From what preceeds, it appears evident that these types of irregularly fenestral, eventually poorly laminated dolomitic build ups result from bacterial reduction of sulfate in presence of organic matter. Such as illustrated in ROUCHY et al (1985, Plate 1, B).

c. A third type a microstructures found in these slope build ups is illustrated on Plate 14. The rock is primarily made of microbial dolomite (10-40 μm) with irregular fenestrae grossly elongated along

Plate 14. A.- Microstructure of a slope microbial build-up (negative photograph); beside open fenestrae in a dolomitic matrix, one recognizes dolomitized scattered traces of originally organic films (arrows); some of these seem to be preserved in their original position such as below arrow 'a' where can be seen two superposed wavy and concordant films and below arrow 'b' which points to a series of stacked films some of which extend to the right. This rock was also probably some sort of laminite originally, or contained layers of stacked organic films. Scale bar : 1 mm. **B.-** Millimetric stromatolite locally developed in the dolomitic mass of Plate 14, A. Scale bar : 2 mm. **C.-** Calcitized laminated gypsum (Zeit) showing brownish organic subcontinuous films and fragments of such films scattered in the central calcitic mosaic; note the drusy aspect of this mosaic showing microsparitic crystals around the organic films and their progressive coarsening toward center of mosaic. Scale bar : 100 μm.

the stratification, and significant intercrystalline porosity. As already reported above, one notes the presence of kerogen between as well as within dolomite crystals. This dolomite contains fragments of horizontal straight or wavy organic films (about 150 μm thick) mineralized or not; at places one can still recognize what could have been the original superposition of these films in particular laminae (see arrows 'a' and 'b' on Plate 14). Mineralization of these films results from an abundance of nucleation sites leading to the development of closely packed small (5-10 μm) dolomitic crystals. In this general framework one can identify small stromatolitic domes, which may be part of an overall altered stromatolitic structure, or represent local swell of laminites (Plate 14, B).

4. Conclusions on stromatolite significance.

The successive lagoonal stromatolitic horizons record a progressive change in environmental conditions from (1) normal open lagoonal waters where stromatolites are associated with a diverse community including red algae (Plate 10), to (2) pure stromatolitic bed, showing an interesting diversification of morphologies and well preserved microstructures, but completely deprived of fossils (except ichnofossils of halophilic organisms on top of stromatolitic horizon), although growing from bioclastic substrate rich in molluscs (Plates 8, 9), to (3) monotonous undulatory stromatolitic bed intercalated between lagoonal sands and silts, and suffering the impact of increased diagenesis, mainly various dolomitization patterns (Plate 11). This last lagoonal stromatolitic level, interpreted as a temporary and localized evaporitic pulse, clearly announces by all its features the upper 30 m. thick stromatolitic blanket (Plate 12; Figure 2). The upper stromatolites on their side, start from a marked discontinuity, develop larger domes and heads and cover the whole carbonate complex. Although we have shown that the building mats were initially soft, the complex shows no dessication cracks and is interpreted as shallow subaquatic. Early microbial dolomitization of films and filaments originated lithification of these stromatolites and their resistance to

Plate 15. A.- Surface of bed made of oncoids plastering locally the talus slope. B.- Thin section of oncoids showing loose cementation at contact points. Scale bar : 2 mm. C.- Silicified nodules in basal lagoonal bed in association with cyanophytes. Scale bar : 200 μm. D.- View of very well preserved and altered (black) coccoid cyanophytes in silica.

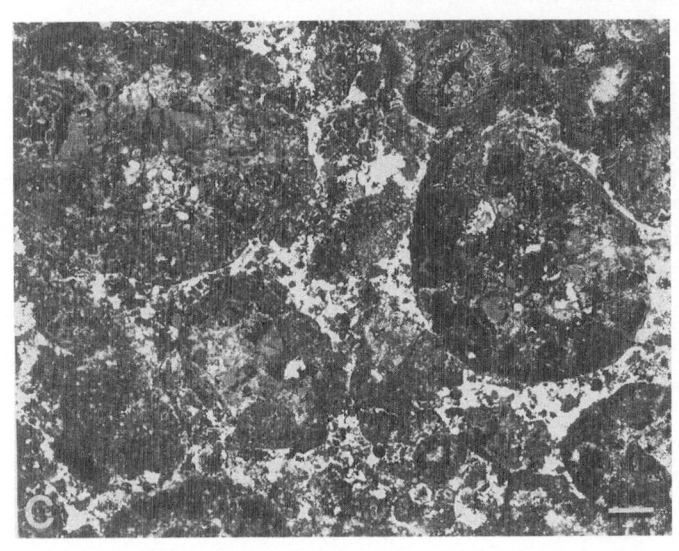

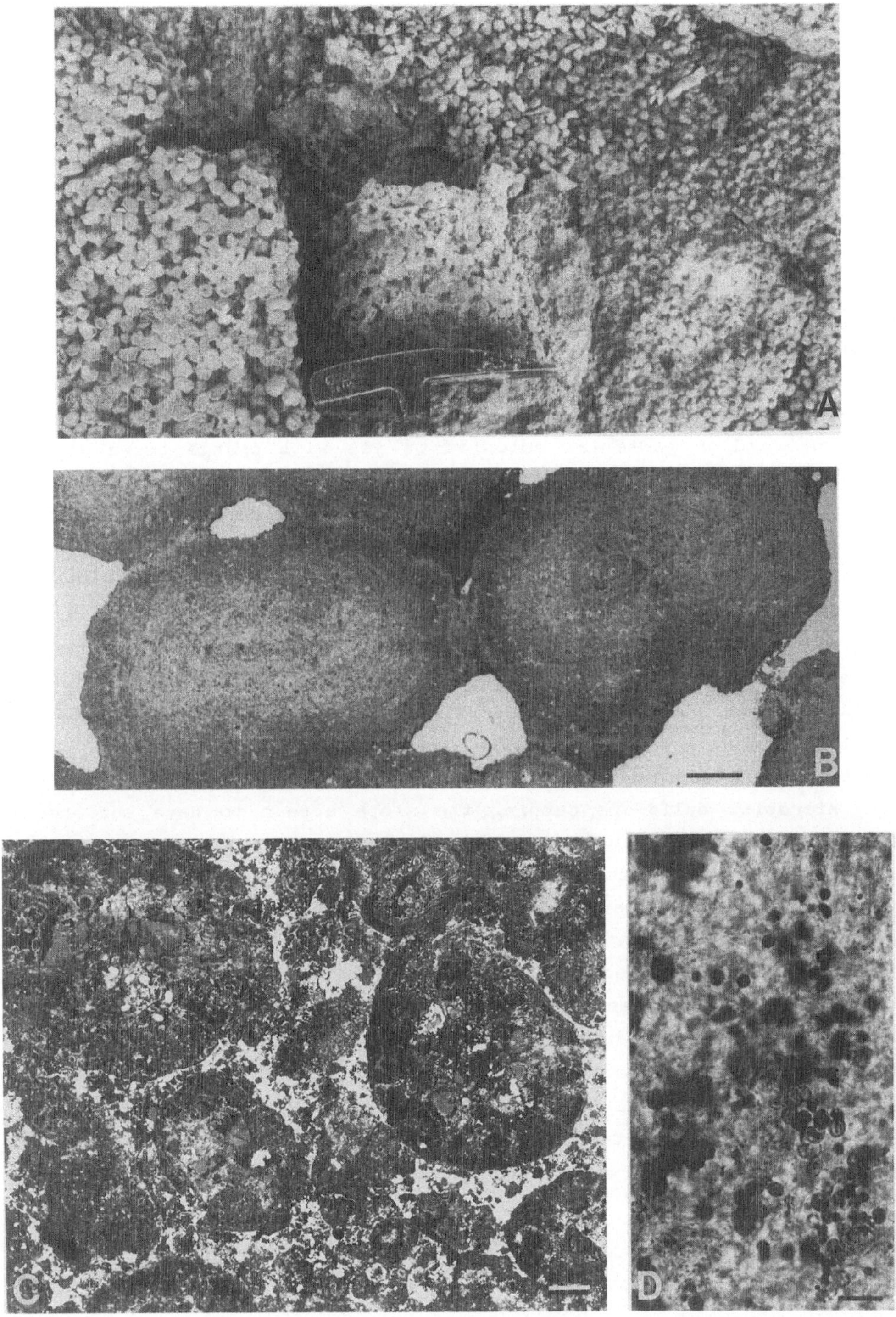

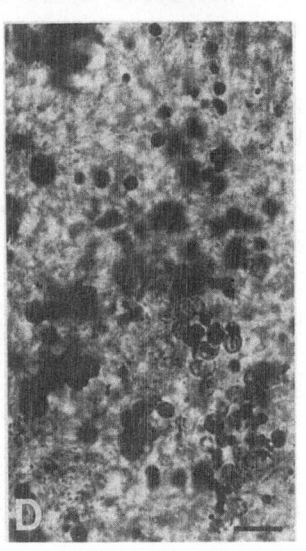

dessication during low water levels. The upper stromatolites thus form
an independant "reefal" (biostromal) microbial unit which postates the
coral build ups and associated facies from lagoon to slope. They
witness drastic environmental changes, namely the end of open marine
stable conditions as a whole and the initiation of highly restricted
environmental settings leading to evaporitic conditions during phases
of low stands; fluctuations of sea level are fossilized by the small
terraces that locally cut the reef front and their later recolonization
by microbial build ups (Plate 7). In this scheme evaporite
precipitation was most probably diachronous between the upper and the
lower parts of the reef front during sea level drops.

Main diagenetic features of upper stromatolite are dominated by
dolomitization processes essentially associated with microbial
filaments and/or bacterial colonies or yet with orange brown organic
matter infilling intercrystalline porosities; such processes do not
seem to be necessarily related to any presently published models, as
observed bacterial dolomite types have been found presently developing
in the vadose zone as well as at depths of several thousand meters in
association with organic laminites or organic matter in process of
maturation (MONTY 1986 a, and internal reports). Other diagenetic
processes include localized silicification in association with
sulfates, and sulfate replacement by carbonates. Gypsum presently
infilling cavities is a late cement probably resulting from
remobilisation processes.

Microbial build-ups capping the slope appear to have suffered a
much stonger diagenesis besides microbial dolomitization. An important
aspect of this diagenesis results from microbial sulfate reduction in
presence of organic laminites and stromatolites particularly rich in
kerogenous organic matter; these reactions led to development of
features very reminiscent of those observed in Quseir and the Zeit (see
above). Some of these build-ups are characterized by significant
development of strontianite (and eventually celestite). These and other
diagenetic features probably result from their much closer proximity of
the evaporitic basin.

Finally the importance of diagenesis in development or alteration
of stromatolitic microstructures should incite us to great care when

Plate 16. A.- Development of flabellate filamentous growths interfering
with the concentric lamination near the outer part of oncoids. Scale
bar : 200 μm. B.- Close up of flabellate growth of filaments. Scale bar
200 μm.

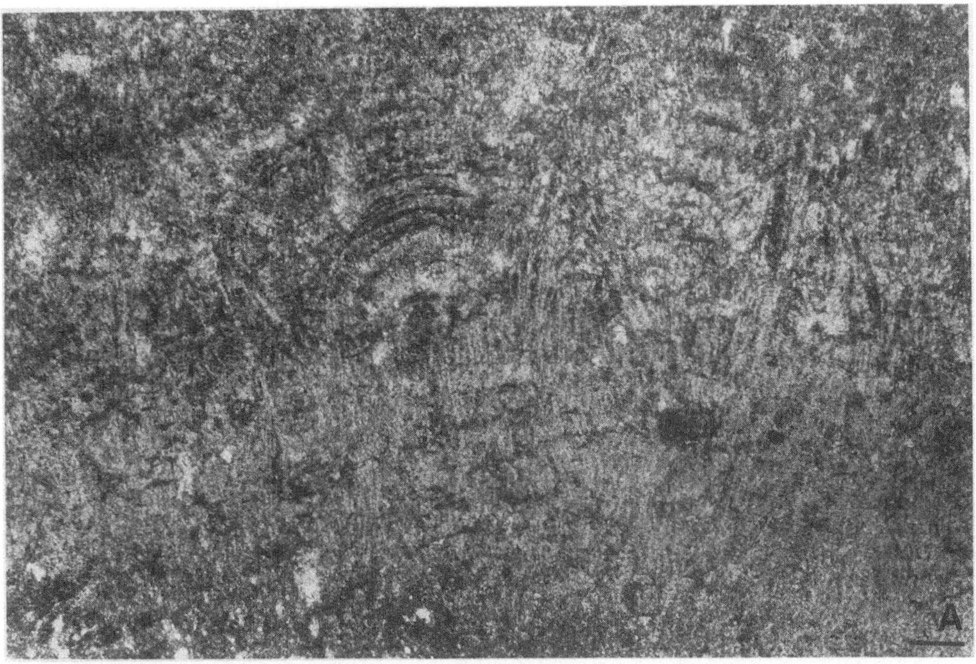

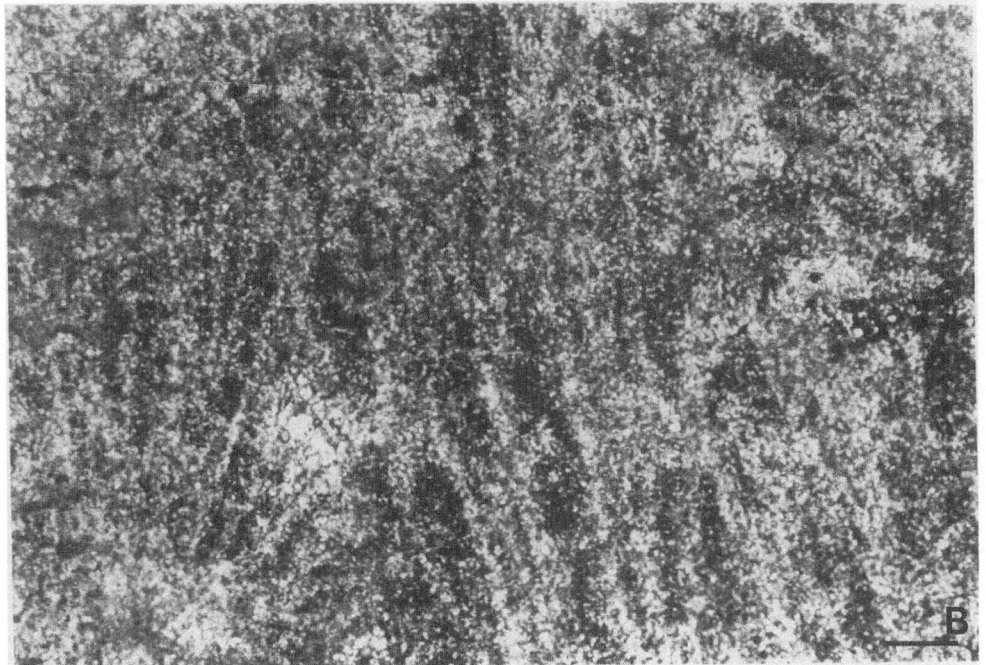

interpreting and characterizing fossils stromatolites.

    5. Oncolites.

    At places, a decimeter-thick oncolitic layer plasters the talus
slope. The oncoids, 1 cm. in average diameter, are closely packed
(Plate 15, A) and poorly cemented by small micritic bridges along their
mutual contacts (Plate 15, B). Actual intergranular porosity is
extremely high; locally, pockets of sediment mainly composed of
peloids, rare benthic foraminifera and various clasts are trapped
between oncoids; they may however be younger infillings. In thin
section, oncoids display a laminated concentric microstructure made of
alternating dense micritic and clear microsparitic laminae; in the
outer half however this pattern is disturbed and/or replaced by
flabellate growths of filaments (Plate 16). At present the
environmental significance and the stabilization of this oncolitic bed
is not clear; eventually, similarities between sediments trapped
between oncoids and those locally associated with the upper
stromatolites may suggest a similar age, i.e. pre evaporitic.
    Complex centimeter sized silicified nodules have also been found
in lagoonal beds (Plate 15, C); they contain a variety of unidentified
grains, although these may evoque grains and fragments found in
lagoonal rhodolites described above (Plate 10) in association with
stromatolites; they could accordingly represent nodules partly cemented
and coated by rhodophytic blades. The cement is also siliceous. This
silica is loaded with perfectly preserved filaments and before all
coccoid cyanobacteria, Plate 15, D; these are well known for the amount
of mucilage their can excrete, mucilage in which we have observed
nucleation of silica (A.M., C.M.). Whether this silicification is
Miocene or more Recent cannot be ascertained.

IV. STROMATOLITES WITH EVAPORITES.

In the basinal deposits, several types of stromatolites or cyanobacterial laminites occur within interevaporitic laminated sediments,
at the base as well as at the top of transitional contacts between
marls and evaporites. At places however stromatolites occur entirely
within crystalline gypsum. The following reported occurrences are

are selected examples and other types may yet occur in outcrop or in subsurface.

A. CEREBROID STROMATOLITES.

This very particular type of stromatolites (Plate 17) has been found paving a distinct level of a marl intercalation between two beds of gypsum at Wadi Feiran (Sinai). They were initially described by MONTY (1980). Recently, similar forms have been reported from intergypsum layers at Gebel Zeit (J. Marie, oral comm.). Similar types were described by BRADLEY (1929) from the eocene Green River Formation; since his description more accessible outcrops were open along the new road from Douglas Pass to the Radar Station. These bowl-shaped stromatolites are 10-30 cm in diameter with an average height of 10 cm (Plate 17,A). The colonies are associated with primary radial ooïds cemented by gypsum. Growth starts from binded sediment, from a crust, a clast or even sulfate debris. It widens rapidly upward by active divergent branching (dendroid type HOFMANN, 1969). This mode of growth leads to the differenciation of a cerebroid morphology on the upper surface. This morphology is reminiscent of the calyx structure described by LOGAN et al (1974) from Shark Bay stromatolites. However, if in this latter case dichotomy of columns is a passive phenomenon due to erosional process, our calyx structure results from active biologically controlled dichotomy as indicated by arrow on Plate 17. pointing changing growth axes; these features, the perfectly preserved morphology and the presence of branches growing almost horizontally require rapid lithification most probably due to bio-dolomitization.

The microstructure appears very monotonous (Plate 17,C): as a whole it is composed of stacked laminae generally very loose at their base, where they show many small subhorizontal fenestrae; laminae become more micritic and denser toward the top and contain binded grains (Plate 17,D). Incorporated grains include some benthic foraminifera, ostracod shells, some glauconitic detrital gypsum grains and ooids; at places the outer layer of radial ooids kept on growing in situ toward center of neighbouring fenestrae (Plate 17,F, arrow). Locally small horizontal fenestrae at the base of each lamina may differentiate vertical tubules (Plate 17,D,E) which may replace vanished bundles of filaments. The whole stromatolitic colony is surrounded by an unlaminated fenestral crust (Plate 17,C) against which abut stromatolitic laminae. This external coating is encrusted by quantities of small tubes (worms). The abundant fenestrae occurring in these stromatolites are coated with an isopachous fibrous aragonitic cement and/or can be infilled with anhydrite. The monotaxonomic encrusting benthos (worm tubes), the presence of detrital gypsum as well as of anhydrite infilling cavities, the abundance of radial aragonitic ooids most of which grew *in situ* (even perhaps in stromatolites and fenestrae), the scarcity of the associated fauna indicate strong ecological stresses bound to hypersaline conditions. Furthermore, the presence of nicely developed radial ooids (300 μm in diameter), the morphology of the colonies and the absence of any preferential growth direction indicate quiet isotropic subaquatic environment.

B. CYANOBACTERIAL LAMINITES WITHIN SELENITE GYPSUM.

The recently worked quarry of Ras Malaab (Sinai), revealed well preserved primary structure of gypsum below the anhydritic alteration crust. Two beds of gypsum are separated by a layer of lime sands (including bivalves coquinas). They present a crystalline selenitic facies with imbricated decimetric crystals. The lower part of the upper gypsum bed ($\approx$ 3,50 m. thick) shows 50 cm. high oriented gypsum crystals, (010) plane in vertical position, displaying a planar and regular laminated pattern. The lamination results from the succession of millimetric to centimetric laminae alternatively rich and poor in filamentous remains (Plate 18,A). The filaments with their sheaths have

Plate17. A.- Oblique view of cerebroid stromatolite lying on oolitic sand (arrow S); note the pustular fenestral crust that surrounds the stromatolite (upper arrow). Photo also shows the central part which always emerges from the calyx. Scale bar : 1 cm. **B.-** Apical view of calyx. Scale in cms. **C.-** Vertical thin section through peripheral stromatolitic branches; the lower third of photograph clearly shows stromatolitic lamination, underlined by sulfate infilled fenestrae, abutting against the non laminated outer fenestral crust, probably build by coccoid cyanobacteria (fc on photo). Scale bar : 5 mm. **D.-** Detail of lamination showing fenestral layers passing upward to dense micritic ones with trapped grains. Scale bar 400 μm. **E.-** Detail of fenestral base of lamination showing vertical tubules which may be molds of erect bundles of filaments. Scale bar : 100 μm (Ph. C.M.). **F.-** Radial ooid keeping on growing in free space of uncemented fenestrae. Scale bar : 100 μm. **G.-** Thin section through cerebroid stromatolite showing active branching following changes in growth axis.

177

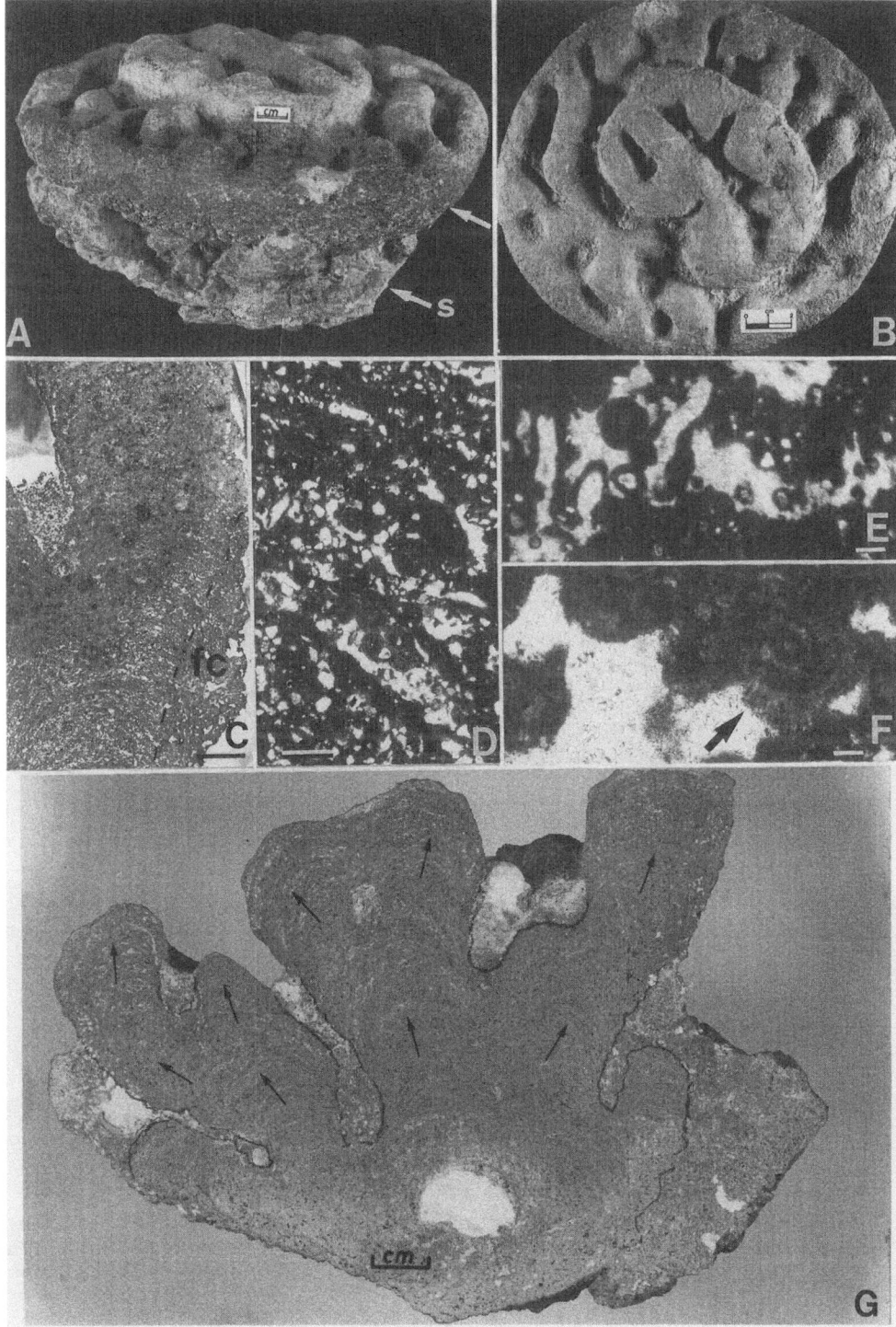

an average diameter of 60 μm (Plate 18, B). In filament-rich laminae, filaments are tangled with a majority of vertically oriented ones. X-Ray diffractometry reveals the monomineralogical character of filaments sheaths (calcitic micrite) infilled with gypsum.

This pattern is very reminiscent of structures described in Messinian gypsum from the Mediterranean Basin by ROUCHY and MONTY (1981). In messinian gypsum, the filaments have been compared with micritized sheaths of *Scytonema* on the base of growth pattern and sheath structure. In the Cenozoic this cyanobacteria presents completely mineralized sheaths (calcitic micrite) in fresh to slightly brackish waters, a process which is inhibited by definitly brackish to marine waters. Significant brackish influences have been identified by the microflora (diatoms) in the messinian strata of the Polemi Basin associated with cyanobacterial-selenite laminites almost identical to, and lying in a sedimentary suite very close to those of Ras Malaab (ROUCHY and MONTY 1981, pp 168-171). On these and other grounds we visualized deposition in restricted areas protected from energetic marine incursion (liable to import bioclasts and grains, totally absent in the two situations) by lumachellic bars. These data lead us to tentatively bind the formation of the Sinai (Ras Malaab) laminites to periodic (seasonal ?) surficial blooms of calcified cyanobacteria of the *Scytonema*-type during short periods of lowered salinities; these should not have altered or induced discontinuities in the underlying selenite crystals if these remained bathed by brines under an eventual halocline beneath the mat; return to hypersaline conditions killed cyanobacterial bloom, reactivated gypsum growth (which, in fact, may have kept on growing and impregnating the base of the mat during its development) and perfectly fossilized the cyanobacterial benthic laminite.

C. MISCELLANEOUS STROMATOLITES AND LAMINITES.

Stromatolites and algal laminites are sporadically distributed near the lower and upper limits of the gypsum beds and, at places, into

Plate 18. A.- Cyanobacterial laminites included into selinitic gypsum (clivage (010) of a large gypsum crystal). Lamination results from alternation of layers rich in cyanobacterial filaments with clear layers almost devoid of filaments. Ras Malaab (Sinai). Scale bar is 1 mm. B.- Close up of cyanobacterial filaments. Scale bar is 500 μm.

the massive gypsum. They form most of the transitional contacts between restricted marine marls and gypsum beds. At least three types of structures have been noticed.

The first type forms scattered decimetric bowl-shaped stromatolites or biscuits overlying gypsum surface in various localities (Gebel Zeit, Abu Shaar el Qibli plain etc..); they are made of contorted millimetric laminites.

The second type consists of thin layers (cm to dm thick) of subplanar to undulated laminites (Plate 19, A) either at the base, at the top or within gypsum.

The internal organization of these two types of microbial deposits consist in an alternation of dense and light layers without preserved filaments. The dark layers contain abundant calcite pseudomorphs after lenticular gypsum witnessing the permanence of hypersaline conditions during the microbial growth (Plate 19, B).

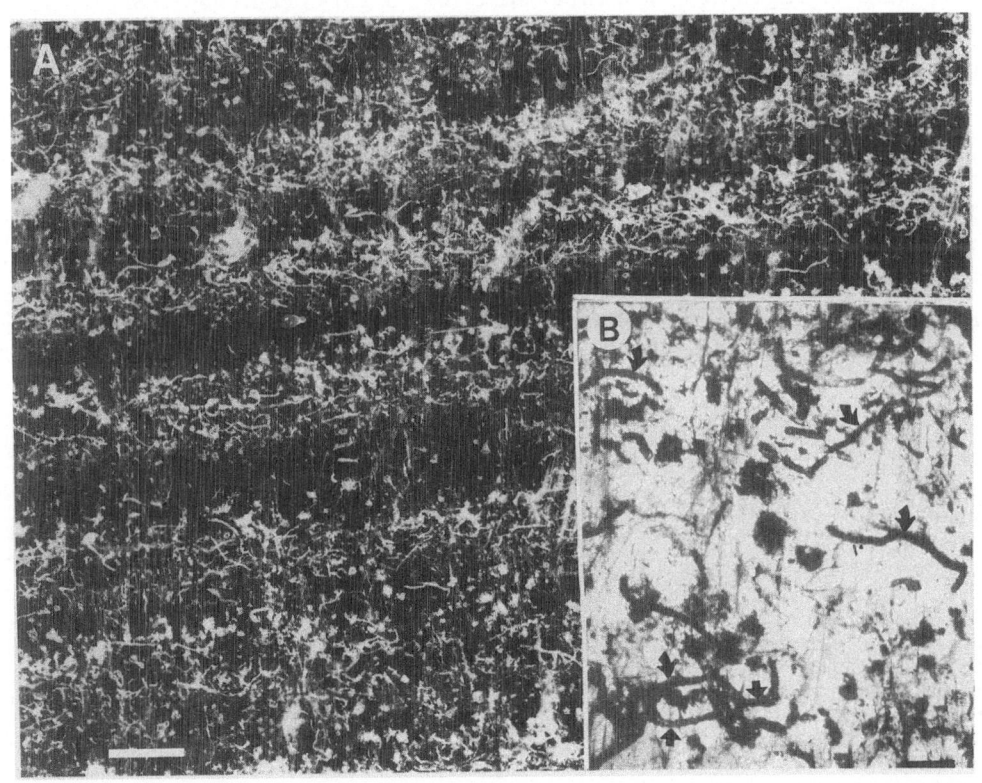

As is shown on Plate 19,A, the laminites can be associated with highly porous diagenetic carbonates (calcite, dolomite, ankerite) irregularly developed at the expenses of over- or underlaying gypsum. These carbonates are considered as resulting from diagenetic processes of bacterial reduction of sulfates (ROUCHY *et al*, 1985; PIERRE and ROUCHY, 1986) in close relationship with organic rich sediments (cyano-bacterial mats, etc..). As reported earlier such biodiagenetic carbonate abound in fenestrae and cavities ranging from millimetric to decimetric or even metric sizes; some of the latter ones may show rather thick infillings by "internal sediments" preceeding late sparry calcites or gypsum (Plate 19,C); the infillings consist of a laminated sediment where thick microsparitic to sparitic laminae, with crystals heavily loaded in inclusions, alternate with thin bituminous films made of dark patches and eventual small filaments; at place they show traces of diffuse iron oxides. Such internal sediments should not be mistaken for normal laminites in this sort of highly confused rocks. They moreover record the periodic development of microbial films within cavities.

The third type shows calcitized finely layered laminite in which the overall laminated pattern and eventual tiny stromatolitic domes, resulting from swelling organic laminae, have been perfectly preserved. Thin section shows however complex petrography, including (1) calcitic layers rich in peloids, micritic inclusions, grains of iron sulfides, (2) micritic to pseudo-micritic dolomitic layers with calcitic microsparite cementing residual microcavities, (3) layers of fibrous aragonitic cement, and (4) microbial films eventually rich in bitumen residues.

Plate 19. A.- Polished slab of planar stromatolite (a) plastering biodiagenetic cavernous carbonates after sulfate (b). Anhydrite relicts can be observed in the latter (c). Little Zeit. Scale bar is 1 cm. **B.**- Thin section of the planar stromatolitic part in A. Lamination is made of dense micritic layers alternating with light layers. Note the abundance of small lenticular gypsum crystals, pseudomorphosed into calcite; scale bar is 1 mm. **C.**- Internal laminated sediment infilling large cavity in biodiagenetic carbonate, lamination comprises thin bituminous films showing filamentous remains (cavity dwelling microbes) alternating with much thicker microsparitic to sparitic layers rich in inclusions. Quseir. Scale bar : 5 mm. **D.**- Calcitized laminite showing rather well preserved lamination including protruding tiny stromatolites on top of photograph. Little Zeit. Scale bar : 2 mm.

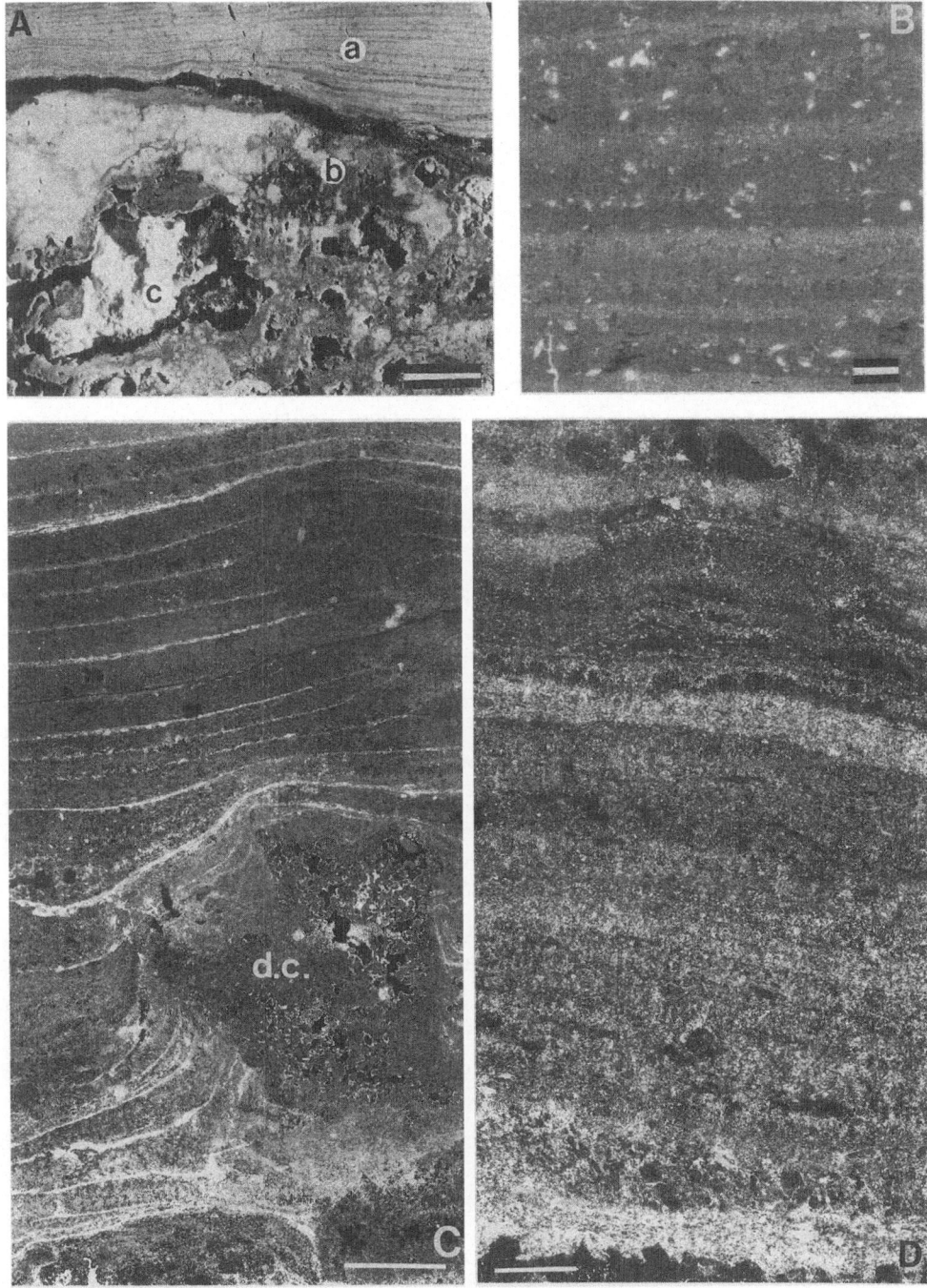

Many other situations could be described yet, but this is not the scope of the paper, the more that we might be driven into extreme details as identification of remains of microbial accretions within sulfates is often considerably hindered by surficial alteration of gypsum to anhydrite, a badly destroying process.

V. CARBONATES-EVAPORITES RELATIONSHIPS. A SYNOPSIS.

The unconformable contact between evaporites pinching out over the base of the reef talus deposits at the foot of the Abu Shaar el Qibli complex (Figure 2) shows that the reef complex predates the deposition of massive evaporites in adjacent grabens. In the South Zeit area, (Little Zeit), the lowermost evaporitic beds overlie either a small coral reef, or a deltaic fan, or yet the preevaporitic *Globigerina* marls. Reef and associated facies formed during a normal open marine stage; this is confirmed by the rich planktonic microfauna of the *Globigerina* marls which, along with the Nukhul Formation, are time equivalent of the fringing reef complex. Nevertheless, beginning of restricted bottom conditions in the basin, with deep, poorly oxygenated waters can be deduced from the antagonism between the rich planktonic thanatocoenoses and the poorly diversified benthic microfaunas (buliminids, CRAVATTE and DUFAURE written comm.). However such "restricted" conditions may have to be considered as fairly gentle at this stage as pointed out by ROUCHY (1982) for the preevaporitic Messinian of the Mediterranean basin where thick diatomitic laminites cannot compare with poorly developed varves in the Red Sea. By the same token relatively well-diversified coral communities and the sharp contact between marine marls and evaporites evidence a rapid, stressing evolution toward evaporitic conditions. Stromatolitic accretions could represent the only witness of a transitional bio-sedimentary phase.

If rapid concentration of marine waters leading to the precipitation of the first evaporites is bound to a drastic drawdown, the presence of evaporitic conditions in shallow lagoons (anhydritic cement and gypsum pseudomorphs in the upper lagoonal stromatolitic horizons) could fit the contemporaneous presence of neighbouring open marine faunas. Increasing salinities and drawdown led first to the death of corals and other strictly marine faunas, then to the exposure of the reef and subaerial leaching.

Estimates of the water depth of the trough adjacent to the Abu Shaar el Qibli complex can be deduced from the altitudinal difference between the top of the reef and the base of the first evaporitic layer in the closest boreholes; this value should be corrected by considering synsedimentary subsidence and tectonics. At present, we can admit that the depth was of several hundred meters, and could have been greater elsewhere in the centralmost part of the rift. Of course, this estimate does not correspond to the depth at which evaporitic deposition took place because of the implied sea level lowering. Nevertheless, we have seen that, even if extremely shallow water conditions and eventual temporary exposures may have occurred in the outskirts of the lagoons and basin, the most important part of evaporitic deposits accumulated subaqueously whatever may have been the water depth. Most of frequent desiccation features occur near the top of the evaporite formation at the time the trough was filling up.

Thus, evaporitic sedimentation, interpreted here as resulting from periods of lowstand, does not implicated overall dessication of the basin, as postulated by the classical paleosabkha depositional models. The important thickness of evaporites reaching 1000 meters at about 1 kilometer away from the carbonate complexes, and several thousand of meters elsewhere in the Gulf, has to be coupled with the various sedimentary and structural features of sulfates. Such a combined analysis yields evidences that a great part of evaporites must have precipitated from increasingly concentrated brines as subaqueous deposits. This led to extreme conditions required by the precipitation of K and Mg salts bound to lower drawdown yet. The evaporitic phase rapidly filled the existing depressions while most of the marginal carbonate complex was exposed.

Environmental interpretation of evaporites is complicated by repeated fluctuations in sea level and water concentrations as is indicated on the edges of the basin by the regular alternation of sulfate layers and marly diatomitic sediments. The fossil content of these diatomitic sediments (foraminifera, calcareous nannoplankton, fishes) are undoubtedly characteristic of marine conditions (ROUCHY 1983, unpublished report; GAUDANT and ROUCHY, in press.; NOEL and ROUCHY in preparation). General features of intraevaporitic sediments, such as the very thin and regular laminated structures of deposits, the scarcity of benthic fauna are all evidences in favor of alternating phases of water-stratification with stages of poorly oxygenated to anoxic lower bodies of water and preevaporitic stage which lacks these features. The sharp contact between laminated diatomitic marls and sulfates suggests that changes in salinity were sudden (overturns ?). The stromatolitic accretions immediately below and above these contacts are the only witnesses of the environmental crisis bound to abrupt changes from marine to hypersaline conditions or vice-versa.

Intra-gypsum stromatolites reveal that algal mats accretions could accomodate an overall evaporitic environment precipitating nanno-crystalline or selenitic gypsum; in the latter case, laminite-building cyanobacteria (Plate 18) could survive in a dormant state (cysts) during phases of hypersalinities to bloom again when suitable conditions settled again. Locally cerebroid stromatolites grew in somewhat isolated coastal bodies of water floored by radial ooids and intermittently submitted to evaporitic conditions; these shallow areas were isolated from more open marine environments floored by restricted benthic forams, glauconite, etc.. .

By the time of the main evaporitic event, repeated fluctuations of sea level produced alternate exposures and flooding of the main reefal complex (Abu Shaar et Qibli build ups) while smaller reefs (Little Zeit) were rapidly overlain by evaporites. Such fluctuations are also probably reponsible for the notches and small terraces cutting the reef front. These changes joint to alteration of physico-chemical parameters progressively exterminated the coral reef ecosystem and opened the way to microbial accretions, first localized (lagoon stromatolites), then incresingly extensive to finally plaster the top and the flanks of the whole carbonate complex with outstanding build-ups. The piling up of stromatolitic biostromes made of globular and/or planar stromatolites constituted a new reef by itself, i.e. a microbial reef. Accordingly as discussed in MONTY (1984, 1986 in press), microbes and particularly stromatolites are the last traces of life or the last bioaccretions to survive environmental crisis. Growth of these upper stromatolites is of course diachronous down the reef slope as well as with that of basinal stromatolites.

Dolomitization is an important feature bound to the related events. We showed that a greater part of it was microbial in origin either because crystals grew around microbial colonies which seem to guide their morphology, or because they nucleate around microbial filaments to form dolomitic chains, or yet because dolomite rhombs and globule form at the expense of kerogenous organic matter which is fairly abundant as films or intercrystalline and intracrystalline patches; stromatolitic organic matter bound to sulfates which were deposited on the reef slope is responsible for the formation of vuggy biodiagenetic build-ups after bacterial sulfate reduction.

In conclusion, the demonstrative quality of the Middle-Miocene outcrops of the Gulf of Suez provides several useful guides for understanding the relationships between three main components of an evaporitic basin : reefs, stromatolites and evaporites. As in other examples of the stratigraphic column (Silurian of Michigan, Devonian of Alberta, Eocene of Spain, Messinian of the Mediterranean, etc...), massive metazoan build-ups (stromatoporoid, corals s.l.) and their associated lagoonal and talus facies predates the beginning of the evaporitic precipitations. Microbial accretions (stromatolites) occur during transitional stages between marine (preevaporitic) and evaporitic conditions when increasing salinity and confinement led to death of corals or any other associated marine organisms.

REFERENCES.

BERNET-ROLLANDE M.C.,MAURIN A.F. and MONTY C.L.,1980.- *Porites* vs Stromatolites at Santa Pola (Spain): a Miocene sedimentological puzzle. 26* Congr. Geol. Int. Paris, 2, sect. 6. 12, p. 434, résumé.

BOUDREAUX A., 1973.- Calcareous nannoplancton ranges. *In*: WHIRTSMARCH R.B. *et al*; IN. Rep. Deep Sea Drill. Proj., 23, Washington (U.S. Governement Office), p. 1073-1090.

BRADLEY W.H., 1929.- Algal reefs and oolites of the Green River Formation. U.S.G.S. Profess. paper, 154-G, p. 203-223.

DAVIES G.R., 1977.- Carbonate-Anhydrite Facies Relationships; Otto Fiord Formation (Mississipian-Pennsylvanian), Canda Artic Archipelo. *In*: FISHER J.H. (ed), Reefs and Evaporites- Concepts and Depositional Models, Studies in Geology n°5, Amer. Assoc. Petrol. Geol.,Tulsa, Oklahoma, p. 145-168.

ESTEBAN M., 1979.- Significance of the Upper Miocene Coral reefs of the Western Mediterranean, Pal., Pal., Pal., 29, p. 169-188.

EL HADDAD A., AISSAOUI D.M. and SOLIMAN S.A., 1983/1984.- Mixed carbonate-siliclastic sedimentation on a Miocene fault block, Gulf of Suez, Egypt. Sediment. Geol., 37, p. 185-202.

EL HEINY I. et MARTIN E., 1981.- Miocene foraminiferal and calcareous nannoplancton assemblages from the Gulf of Suez region and correlation. Géol. Méd., VIII, p. 101-108.

GAUDANT J. and ROUCHY J.M., in preparation.- Ras Dîb: un nouveau gisement de poissons fossiles du Miocen de Gebel Zeit (Egypte).

GHORAB M.A., EL SHAZLY E.M., ABDEL GAWAD A., MORSHED T., AMMAR A.A. and IBRAHIM A.M., 1969,- Discovery of potassium salts in the evaporites of some oil wells in the Gulf of Suez region (abs.). Geol. Soc. U.A.R., 7ᵗʰ Ann. Mtg Abstracts, Cairo, sess. 1, p.3.4.

GREGORY J.W., 1906.- On a Collection of Fossil Corals from Eastern Egypt, Abu Roash, and Sinai, v, 3, p. 50-58.

HASSAN F. and EL DASHLOUTY S., 1970.- Miocene evaporites of the Gulf of Suez region and their significance. Am. Assoc. Petrol. Geol. Bull., 54, p. 1686-1696.

HOFMANN H.J. 1969.- Attributes of stromatolites. Geol. Survey of Canada, paper 69-39, 58 p.

HUH J.M., BRIGGS L.I. and GILL D., 1977.- Depositional Environments of Pinnacle Reefs, Niagara and Salina Groups, Northern Shelf, Michigan Basin. *In*: FISHER J.H. (ed), Reefs and Evaporites- Concepts and Depositional Models, Studies in Geology n°5, Amer. Assoc. Petrol. Geol.,Tulsa, Oklahoma, p. 1-22.

KERDANY M.T., 1968.- Note on the planktonic zonation of the Miocene in the Gulf of Suez region, U.A.R. Comm. Med. Neog.strat., Proc. IV sess;, Bologna, 1967, Giorn. di Geol., ser. 29, 35, p. 157-166.

KIRKLAND D.W. and EVANS R., 1976.- Origin of limestone buttes, Gypsum Plain, Culberson County, Texas. Am. Ass. Pet. Geol. Bull., V. 60, p. 2005-2018.

KIRKLAND D.W. and EVANS R., 1973.- Marine Evaporites. Origin, Diagenesis and Geochemistry. Benchonark Papers in Geol., Dowden, Hutchison and Ross, Inc. 426 p.

LOGAN B.W., HOFFMAN P., and GEBELEIN C.D.,1974.- Algal Mats, Cryptalgal Fabrics, and Structures, Hamelin Pool, Western Australia. *In* LOGAN B.W. *et al*, Evolution and Diagenesis of Quaternary Carbonate Sequences, Shark Bay, Western Australia. Amer. Assoc. Petrol. Geol., Mem. 22, p. 140-194.

MADGWICK T.G., MOON S.W. and SADE K.H., 1920.- Preliminary geological report on the Abu Shaar el Qibli (Black Hill) district. Petrol. Research Bull. 6, Governement press, Cairo, 11 p.

MAIOLA R.J. and GLOVER E.D., 1965.- Recent anhydrite from CLAYTON Playa, Nevada. Am. Mineralogist, 50, 11-12, p. 2063-2069.

MESOLELLA K.J., ROBINSON J.D., MC CORNICK L.N. and ORMISTON A.R., 1974.- Cyclic Deposition of Silurian Carbonates and Evaporites in Michigan Basin. Amer. Assoc. Petrol. Geol. Bull., 58, p. 34-62.

MONTY C.L.V., 1976.- The origin and development of cryptalgal fabrics, *In*: Stromatolites, WALTER M. (ed.), Elsevier Scientific Publ., Dev. in Sedim 20, p. 193-249.

MONTY C.L.V., 1982a.- Microbial Spars. In: Intern. Assoc. of Sediment., Abstracts of paper, 11th Intern. Congress on Sediment, Hamilton, Ontario, Canada, p. 26.

MONTY C.L.V., 1982b.- Cavity or fissure dwelling stromatolites (endostromatolites) from Belgian Devonian mud mounds (extended abstract). Ann. Soc. géol. Belg., 105(2), p. 343-344.

MONTY C.L.V., 1984.- Stromatolites in earth history. Terra Cognita, 4, n°4, p. 423-430.

MONTY C.L.V., 1986 a.- Microbial dolomites. Abstract. 12th Intern. Sedim. Congress. Canberra. Australia.

MONTY C.L.V., 1986 b.- Range and significance of cavity-dwelling or endostromatolites. Abstract. 12th Intern. Sedim. Congress. Canberra. Australia.

MONTY C.L.V., in press.- Interactions événements géologiques-stromatolites, et vice-versa. Interactions geological events-stromatolites and vice-versa. Bull. Cent. Rech. Explo.- Prod. Elf. Aquitaine.

MONTY C.L.V. & MAURIN A., 1982.- Microbial accretions and cavity dwelling stromatolites in reefs and mounds. In: Intern. Assoc. of Sediment., Abstracts of paper, 11th Intern. Congress on Sediment, Hamilton, Ontario, Canada, p. 28.

ORTI-CABI F., PUEYO-MUR J.J. and ROSELL-ORTIZ L., 1984.- Some aspects of the evaporite deposition in the Upper Eocene South-Pyrenean potash basin (Spain). 5th Europ. Congr sediment., Marseille, Abstract.

PIERRE C. and ROUCHY J.M., 1986.- Sedimentological and geochemical diagnostic features of carbonates replacements after sulfate evaporites: application to the example of the Middle Miocene of Egypt. Int. Congr. Geochemistry of the Earth surf. and processes of mineral formation, Granada, (Spain), Abstract, p. 28-29.

ROUCHY J.M., 1979.- La sédimentation évaporitique sur les marge messiniennes. 7th Int. Congr. on Medit. Neogene, Athènes 1979. Ann. Geol. des Pays hellen., Athènes, h.s., 3; p. 1051-1061.

ROUCHY J.M., 1982.- La génese des evaporites messiniennes de Méditerranée. Mém. Mus. nat. Hist. nat., Paris, 267 p.

ROUCHY J.M., BERNET-ROLLANDE M.C. and MAURIN A.F., 1986.- Methodes de description, terrain, subsurface, laboratoire. In: Caractérisation des séries à évaporites en exploration pétrolière. Tome 1, Géologie, Chap. 3, in press, Technip

ROUCHY J.M., BERNET-ROLLANDE M.C., MAURIN A.F. and MONTY C.L.V., 1983.- Signification sédimentologique et paléontologique des divers types de carbonates bioconstruits associées aux evaporites du Miocène moyen près du Gebel Esh Mellaha (Egypte). C. R. Acad. Sci., Paris, sér.II, 296, p. 457-462.

ROUCHY J.M. and MONTY C.L., 1981.- Stromatolites and cryptalgal laminites associated with Messinian gypsum of Cyprus. *In*: MONTY C. (ed.), Phanerozoic Stromatolites. Case Histories. Springer Verlag, Berlin, Heidelberg, New-York, p. 155-180.

ROUCHY J.M., MONTY C., BERNET-ROLLANDE M.C. and MAURIN A.F., 1982.- Mid Miocene stromatolites of Gebel Esh Mellaha (Egypt) and their sedimentological interest, *In*: Intern. Assoc. of Sediment., Abstracts of paper, 11[tr] Intern. Congress on Sediment, Hamilton, Ontario, Canada,

ROUCHY J.M., MONTY C., PIERRE C., BERNET-ROLLANDE M.C., MAURIN A.F and PERTHUISOT J.P., 1985.- Genèse des corps carbonatés diagénétiques par réductions de sulfates dans le Miocene évaporitique du Golfe de Suez et de la Mer Rouge. C. R. Acad. Sci., Paris, sér.II, 301, p. 1193-1198.

ROUCHY J.M., SAINT MARTIN J.P., BERNET-ROLLANDE M.C. and MAURIN A.F., - Evolution et antagonisme des constructions animales et végétales à la fin du Miocene en Méditerranée, biologie et sédimentologie. in preparation.

SCHREIBER B.C.,1978.- Environments of subaqueous gypsum deposition. *In*: DEAN W.E. and SCHREIBER B.C. (eds.), Marine Evaporites. Soc. Econ. Paleont. Miner., Short Course, 4, p. 43-73.

SELLWOOD B.W. and NETHERWOOD R.E., 1984.- Facies evolution in the Gulf of Suez area: sedimentation history as an indicator of rift initiation and development. Modern Geol., 9, p.43-69.

SHEARMAN D.J., 1971.- Marine evaporites. The calcium sulfate facies. Unpublished notebook, the Univ. of Calgary, Assoc. Sed. Petrol. Geol. Seminar, 65 p.

SHEARMAN D.J. and ORTI-CABO F., 1978.- Upper Miocene gypsum. San Miguel de Salinas, S.E. Spain. *In*: CATALANO R. *et al* (eds.). Messinian evaporites in the Mediterranean. Mem. Soc. geol. Ital., 1976, XVI, p. 327-339.

VAN LAER P. and MONTY C., 1984.- The cementation of mud mound cavities by microbial spars. 5[th]European Regional Meeting of Sedimentology, I.A.S., Marseille, Abstract of paper.

YOUSSEF E.A.A., 1986.- Depositional and diagenetic models of some Miocene evaporites on the Red Sea Coast, Egypt. Sedimentary Geol., 48, p. 17-36.